NOV 20 1985

TECHNOLOGY, INSTITUTIONS AND GOVERNMENT POLICIES

TECHNOLOGY, INSTITUTIONS AND GOVERNMENT POLICIES

Edited by
Jeffrey James
and
Susumu Watanabe

St. Martin's Press New York

© International Labour Organisation 1985

All rights reserved. For information, write:
St. Martin's Press, Inc., 175 Fifth Avenue, New York, NY 10010
Printed in Hong Kong
Published in the United Kingdom by The Macmillan Press Ltd.
First published in the United States of America in 1985

ISBN 0–312–79006–6

Library of Congress Cataloging in Publication Data

Main entry under title:

Technology, institutions and government policies.

    Includes index.
    1. Technology and state—Addresses, essays,
lectures.   2. Technology and state—Developing
countries—Addresses, essays, lectures.   I. James,
Jeffrey. II. Watanabe, Susumu.
T49.5.T4445   1985   338.4'76   84–17955
ISBN 0–312–79006–6

Contents

Preface

Government policies have an important influence, both positive and negative, on the choice of techniques. A government can, for example, increase the supply of appropriate technologies by encouraging their importation and/or development and related information services. It can also make such technologies more attractive to entrepreneurs by raising their relative profitability in comparison with other alternatives. Even more important is the role of government policies in providing a socio-political and institutional climate, as well as an industrial structure, more favourable to the adoption and spread of such technologies.

This volume, edited by James and Watanabe and financed by a research grant from the Swedish Government, is largely an outcome of a project on the impact of government policies on technology decisions which was undertaken within the framework of the ILO World Employment Programme. The volume contains conceptual and empirical essays which ingeniously identify major constraints on the development, adoption and diffusion of 'appropriate technologies' and analyse government measures which influence entrepreneurial decisions in respect of technology choice. Earlier work on the subject has focused much less on institutional constraints. Indeed, studies on the impact of government policies are few and far between. A very ambitious study to date was the one undertaken by the International Development Research Center (IDRC) of Canada. It covered ten developing countries over a period of three years. While the focus of this project was largely sectoral, the studies in the present volume are far more macroeconomic, with a sharper focus on the role and significance of policies on technological decisions at the micro level. The political economy view of policy choices discussed by Enos, James and Stewart also distinguishes this volume from the others.

Although specific government policies have narrow, well-defined aims, in general the ultimate goal of a technology policy needs to be the building up of national technological capability.

There is as yet little in the nature of a systematic body of empirical knowledge to determine the ingredients of national capacity building. Whatever the essential elements of capacity (e.g. skill formation, institution building, research and development, engineering and management consultancy services), it is clear that building up of this capacity, especially in the least developed countries, can take place only in the long run. Opinions vary a great deal on the means to achieve this capacity. Nevertheless, there is broad consensus among economists and policy-makers that an indigenous technological capacity is essential for the economic development of the Third World.

In its future work, the ILO Technology Programme intends to explore the concept and policy aspects of capacity building. The present volume sets the stage for this work, since governments have a major role to play in strengthening national capacities.

A.S. BHALLA
Chief
Technology and Employment Branch
International Labour Office

Notes on the Contributors

Henry Bruton is Professor of Economics at Williams College. He has served in an advisory capacity in many developing countries and has taught in the University of Bombay. He is author of *Principles of Development Economics* and several other books, and numerous articles in the field of development, productivity and employment in developing nations.

John L. Enos is a Fellow of Magdalen College, Oxford, where he teaches economics. He has worked in India, Turkey and Chile. His publications include *Petroleum, Progress and Profits*, *Planning Development* (with Keith Griffin), and *The Absorption and Diffusion of Imported Technology: The Case of Korea* (editor with W.H. Park).

David J.C. Forsyth is Reader of Economics at the University of Strathclyde. He has been consultant to the ILO and other international organisations. His publications include *US Investment in Scotland* and *Choice of Manufacturing Technology in Sugar Production in Less Developed Countries*, as well as many articles on technology and development of Third World countries.

Jeffrey James is Assistant Professor of Economics at Boston University. He was previously Research Fellow in Development Economics at Queen Elizabeth House, Oxford, and Lecturer in Economics at New College, Oxford. He is co-author of *Transition to Egalitarian Development* (with Keith Griffin), co-editor of *The Economics of New Technology in Developing Countries* (with Frances Stewart), and author of *Consumer Choice in the Third World*.

Frances Stewart is a Fellow of Somerville College and Senior Research Officer at the Institute of Commonwealth Studies, Oxford. She has been a consultant to many organisations including the ILO, UNCTAD and the World Bank. Her work includes *Technology and*

Underdevelopment, Framework for International Financial Cooperation (with A. Sengupta), *Economics of New Technology in Developing Countries* (editor with Jeffrey James), and *Work, Income and Inequality* (editor).

Susumu Watanabe is Research Coordinator in the Technology and Employment Branch of the ILO. He has published many articles on employment, technology and industrialisation. He is editor and main author of *International Sub-Contracting: A Tool of Technology Transfer* and *Technology, Marketing and Industrialisation*.

John White is with the Development Cooperation Directorate of the OECD. A career in journalism with *The Times* was followed by teaching and research at the University of Sussex. His writing on foreign affairs includes a book, *The Politics of Foreign Aid*.

1 Introduction

JEFFREY JAMES and
SUSUMU WATANABE

For over a decade, technology has been the subject of intense debate among those concerned with problems of poverty and unemployment in developing countries. Occasioned mainly by the observation that the technology imported by these countries tends to be inordinately capital intensive, and in other respects inappropriate to their conditions and factor endowments, the debate has centred on three main areas.[1] In the first of these, investigators have sought to understand why such technologies are actually chosen by producers in the Third World, where, despite high rates of growth of industrial production, unemployment and underemployment remain widespread. Secondly, interest has focused on the advocacy of alternative, more appropriate technologies and the precise meaning that is to be given to the notion of 'appropriateness'. Finally, the extent of availability of supply of these alternatives has been examined for a variety of different industries. In each of these areas – and perhaps especially the first and last – progress has been considerable and a number of influential early views have had to be modified or rejected. For instance, in contrast to the once popular idea that technological choice is governed by essentially fixed coefficients of production, we now know that in many industries a fairly wide range of efficient technologies does in fact exist. Moreover, our view of the factors determining the choice from among these alternatives has undergone substantial modification. In particular, the assumptions of the neoclassical model of the firm, which underlies much of the early literature, have come under increasingly severe attack and this critique has demanded a re-examination of the impact of government policies on technology decisions in developing countries.[2] Already, the re-examination has

1

yielded a much richer view of this relationship than that based on the simple neoclassical theory.

Important as these and other recent advances undoubtedly are, they have nevertheless been confined almost entirely to the realm of microeconomics. Seldom have authors addressed themselves to the macro aspects of technology policy or the interactions between the macroeconomic environment and micro decisions for technology.[3] This comparative neglect of macro questions has made it understandably difficult for decision-makers to incorporate technology issues in planning at the national level. This volume, consisting of conceptual and empirical parts, is an attempt to bridge this gap. In varying degrees, the essays also contribute to the major elements of the ongoing appropriate technology debate that was described in the previous paragraph.

CONCEPTUAL APPROACHES

Micro versus Macro Approaches

The essays by Stewart and Bruton in the volume provide a framework which ought to simplify the task of identifying macro policies and institutions that will promote more appropriate technologies at the national level.

Stewart's essay is mainly taxonomic – it seeks to classify general government policies according to their impact on technology decisions at the micro level. Bruton, on the other hand, is concerned with defining the conditions in the economy that will lead to the emergence of what he calls a 'national technology', a broad term which embodies the notion that 'technological advance must occur in response to the conditions within the country's economic and social system rather than be imposed or simply made available from outside sources'. Thus, whereas Stewart focuses primarily on the more conventional, static question of technology choice at a point of time, Bruton deals with a more dynamic type of macroeconomic problem. We shall argue below that these aspects of technology should not be considered independently of each other; rather, they should be viewed as essential complements in an integrated set of macro policies that are designed to secure the adoption of more appropriate technologies.

Stewart's approach begins with a micro model in which technology choice is a function of four variables, namely the firm's objectives, the resources available to it (and the prices it has to pay for these resources), the nature of the market it faces, and the knowledge it possesses about available technological alternatives. Different micro units (for example, small family-owned firms as against large multinational corporations) are likely to differ with respect to at least some and probably all these variables. 'It follows that macro policies may affect micro choice of any given unit by affecting any of these four variables. In addition, the government may affect the balance of choice in the economy as a whole by altering the proportion of resources controlled by various units.' Much of her discussion is concerned to illustrate how more appropriate technological choices at the national level can be achieved by macro policies that are designed, first, to improve the choices of each particular group of micro units and, secondly, to alter the composition of units in favour of those whose choices are relatively labour intensive.

But even if government policies in these two areas do succeed in altering the balance of choice in the economy in a direction which is favourable to appropriate technology, Stewart reminds us at the end of her essay about the worrying fact that techniques are continually changing, and that these changes (which originate for the most part in the industrialised countries) are likely to shift the technology shelf in a generally labour-saving direction. Over time, therefore, the effect of improved choices on the overall degree of capital intensity has to be set against the changes in the shelf itself. That the latter effect is capable of overwhelming the former, can be shown with reference to data from Tanzania.

From experience with a variety of different textile technologies, it had become clear to policy-makers in that country that the firm with the simplest technology not only had a higher trading surplus than any of the other firms, but it had also accumulated far more learning effects. However, when this technology was actually sought by the authorities, they were informed that it had become obsolete and that production had ceased. Of course, older vintages do not always become obsolescent, but at least in Tanzania the problem of unavailability does not appear to have been confined only to the textile industry. On the contrary, 'One reason that the slightly better ability to choose has not altered the shift towards higher technology is in many cases that about the time that one has created a better ability to choose, one can no longer buy the thing then seen to be appropriate.'[4]

One major reason for emphasising the need for a 'national technology' or an 'indigenous technological capability' is that it may avoid this sort of problem. For to the extent that this capability leads to the creation of a stream of efficient, capital-saving technologies, more appropriate choices at the static level will be *reinforced* rather than *undermined* by the dynamic changes in technology.[5]

Bruton begins his search for the conditions that will lead to the emergence of an indigenous technological capability with an 'analysis of the individual enterprise'. Using a micro model considerably different from the traditional neoclassical approach (as too does Stewart), he shows that the first condition for the establishment of an indigenous capability lies on the demand side. In particular, 'Initiation of search for new knowledge by the user is the first step in creating a continuous flow of new technologies over broad areas of the economy.' Bruton suggests that the fulfilment of the demand condition should be encouraged by a set of fiscal policies that induce 'economic agents to do what they would not otherwise do'. For example, if agricultural prices are related positively to yield per unit of land, this may help to induce search for machines whose use facilitates an increase in yield per land unit.

The second condition is that there be an adequate supply response to meet the induced demand for new knowledge. In Bruton's view a domestic capital goods sector is essential to the adequacy of the supply response. In arriving at this conclusion – which, as he admits, runs counter to the static comparative cost doctrine – he stresses the learning opportunities that this sector can provide and the need for a close symbiotic type of relationship between firms that create technology and those that use it.

Bruton's approach raises a number of important and controversial questions about the concept of indigenous technological capability and how government policies may most effectively be used to promote it. How much protection, for example, should be afforded to domestic capital goods producers? And if there is a general case for protecting learning effects in other industries, which among them would yield the highest net social benefits? To what extent will an indigenous technological capability, even if it is realised, lead to the desired pattern of technical change?

Unlike these conceptual essays, the paper by Enos deals with microeconomic aspects of the relationship between technology, institutions and government policies. In proposing game-theory as an alternative to traditional neoclassical micro theory, he takes the debate on choice of technology in developing countries a good deal further, for in at least two fundamental respects the assumptions of the former will often correspond to reality much more closely than those of the latter.

First, Enos' approach emphasises that the choice of technology should be described as a process which 'extends through time, taking at the least a few years, and more likely a few decades, to come to completion'. To anyone who has observed the way in which technology decisions are actually made in developing countries, it is evident that this description is far more realistic than the assumption, implicit in the traditional theory, that such decisions are taken instantaneously. Secondly, whereas in traditional theory it is a single economic agent who makes the relevant choices, game-theory allows as many players as are thought to have an influence to be directly incorporated into the model. In some cases (such as in small firms), of course, technology decisions are made by only one individual and there is no difficulty from this standpoint in applying traditional theory. But in many other cases, when numerous individuals and institutions are involved, any realistic description of the outcome needs to take into account the relative power of each of the participants.

Enos uses a game-theoretic model which includes three players to explain 'the overly capital-intensive' choice that is so typical of firms in developing countries. In this model, the government, represented by the relevant ministry, interacts with the firm owner and the aid donor. But as a general rule, direct government intervention will be more likely to occur in the case of state-owned enterprises than those that are privately owned. In either case though, the Enos approach offers a particularly powerful new tool of analysis of *direct* interventions by government, which, according to Stewart's classification, is the one main respect in which government policies affect technology decisions (the other type of intervention includes all those cases in which the impact of policy is *indirect*).

In the final conceptual essay, James raises an aspect of the relationship between technology and government policies that is not dealt with in any of the three previous essays. He argues that from a policy point of view, it is not enough to know how to alter the balance of technology in the economy in the desired direction. For even an altered technological structure is only *instrumental* to achieving particular goals of government, and one is bound to ask, therefore, whether the goals cannot be more efficiently met through policies that do *not* operate through the technological medium. If appropriate technology is defined with specific reference to the goals of reducing inequality and eliminating poverty (as in Stewart's 'specific characteristics' definition), the question becomes one of assessing the role that policies for appropriate technology ought to play in a redistributive development strategy.

In his assessment, James finds that the strength of the claim for inclusion of appropriate technology policies in this type of strategy depends on both micro- and macroeconomic factors. On the one hand, he shows that it is necessary to establish the microeconomic impact of the policies on the heterogeneous groups among the poor, and on the other, he argues that much depends on the nature of the overall redistributive policy that is pursued by government – whether it is radical and designed to achieve a large-scale redistribution in a short period of time, or whether it is more gradualist.

Political Economy Factors

Political economy factors figure prominently in the essays by Stewart, Enos and James.

Stewart is critical of both the 'first-best' policies often proposed by economists that totally ignore the constraints imposed by political economy factors, as well as the opposite view that political economy factors impose limitations so severe as to rule out *any* scope for the choice of alternative technologies. She argues instead that the feasibility of different macro policies for appropriate technology should be assessed on a case-by-case basis. For what is feasible in one sociopolitical context may be totally unrealistic in another. As a method of assessing the feasibility of various policies Stewart advocates the drawing up of a costs–benefits matrix which shows the gains and losses to the various groups that are affected. In some fortuitous circumstances, policies will pass the Pareto criterion – that there be no losers – but in the vast majority of cases a conflict situation will emerge, in which some groups gain and others lose (and the outcome then depends on the relative strength of the two groups).

The political economy approach that is suggested by Stewart resembles, and is indeed to some extent inherent in, the game-theoretic approach proposed by Enos. For the latter not only incorporates the possibility that numerous agents are involved in technology decisions and that their interests may conflict, but it also embodies a method of analysing how these conflicts are resolved.

For James, the political feasibility of policies for appropriate technology is one criterion for inclusion of these, as opposed to other policies, in a redistributive development strategy. He emphasises that this dimension of technology policies is more relevant to a gradualist than a radical strategy because in the latter political circumstances are likely to be such that almost all policies are feasible.

EMPIRICAL STUDIES

Although they were prepared independently and on different occasions, the empirical studies in Part II are fairly closely linked with the conceptual studies in Part I: Forsyth's with Stewart's, as they both deal with a wide spectrum of macro policies; White's with Enos', since both are concerned with conflicts of multiple objectives pursued by plural agents involved in the choice of technology; and Watanabe's with Bruton's, since they both explore issues of developing indigenous technological capacity in Third World countries.

Forsyth begins his study by asserting that 'an ever-widening range of government activities can be seen to be relevant to, and, indeed, to comprise, technology policy, though the majority of such activities are not viewed in this light by the authorities, and hence constitute implicit rather than explicit policies', and attempts to assess how macroeconomic policies influence the capital intensity of the technology, his major criterion of appropriateness. His preliminary review of literature and cross-country statistical analysis suggest that the scale of production is the key determinant of the capital intensity of industries, while competitive pressure and market power have little or no influence. This might be interpreted as a sign of limited scope for governmental intervention. Forsyth, however, argues that non-scale factors may be important in different ways at different levels of scale, although they are not revealed through standard regression analysis based on sector-wide or industry-wide data. He attempts to explore the issue through a survey of Egyptian firms.

His survey findings indicate that most medium- to large-scale firms in both the private and public sectors tend to opt for the highest feasible degree of sophistication, while small firms work with labour-intensive, often obsolete technology. In Egypt, as in many field studies in different parts of the world, competition among larger firms in domestic and export markets is concerned with product quality and this works as a powerful force promoting the use of high, capital-intensive technology. In the small sectors, in contrast, price competition tends to encourage use of labour-intensive technologies, often of obsolete kinds.

This dual industrial structure has been, in Forsyth's view, largely a result of explicit and implicit government policies. Under the 'Open Door' policy introduced in the early 1970s, the Egyptian Government's explicit technology policy has emphasised the ability to increase foreign exchange earnings and absorption of the most up-to-date technology. Consequently, 'the process of investing in plant and machinery and of creating jobs have not been seen by the government as inextricably intertwined, but rather as separate issues . . . the possibility of generating employment in a large scale by manipulating technology does not seem to have been considered, or, at any rate, given a high priority.'

The government's 'implicit' policies concerning wages, prices, foreign exchange rates, quality standards, taxes and import tariffs, supply of investment funds and equipment, etc., tend to intensify the capital-intensive bias in the medium- and large-scale sectors of Egyptian industries.

Forsyth feels that a reversal of the government's policies might have 'important consequences for labour absorption in many industries', partly through an increased supply of resources to more labour-intensive small sectors. At the same time, he is also aware of the constraints imposed by 'political economy' and is not overly sanguine about the feasibility of any such reversal.

In fact, his own findings suggest obstacles within the small sectors themselves: 'in general no objective preference for labour-intensive technologies was expressed by such firms; on the contrary, small manufacturers using primitive or obsolete equipment by and large wished to re-equip with modern machinery as and when circumstances permitted.' Here we find a case of conflicting criteria in choosing a characteristics-specific definition of 'appropriate technology'.

Another important issue which emerges from Forsyth's findings is related to the role of public sector enterprises in employment creation. Until the early 1970s almost all industrial investment in Egypt took place in the public sector. Even today, 75 per cent of the nation's investment takes place in the same sector. This means that the leading selectors of technology have been public enterprises. As already noted, their main criterion in the choice of technology seems to have been modernity. Employment opportunities are created instead through overmanning of plants and offices. A promising extension of Forsyth's work would seem to lie in the determination of how far the capital-intensive choices that he describes can be attributed to public ownership *per se*. For it may turn out that comparable (in terms of type of product, size, etc.) privately owned firms make similar choices, and hence that the problem does not lie in the specifically public character of Egypt's industrial sector.[6]

White's study deals with the problem of conflicts among multiple objectives pursued by multiple decision-makers involved in a single aid project. Of course, conflicts among 'players' will not always arise. In some cases there will be a consensus with respect to the technology that is to be adopted. White's paper, for example, alludes to the case of a World Bank project in Bangladesh in which

> for shortage of time and information, the Bank's appraisal mission chose the technology that seemed to fit their perception of what was needed, met their organisation requirements, and also had a satisfactory economic justification. As far as the government officials were concerned, aid was available only for medium-cost wells, and this technology conformed with their institutional requirements and perhaps their personal preferences as well. . . In the actual decision making such factors as risk avoidance, appearance of modernity, established procedures, familiar techniques, and by no means least, control, outweighed development policy objectives.

In other cases, however, there will be conflicts between the different parties and a compromise result will be reached. This appears to have been the case, for example, in the choice of technology for a bakery in Tanzania. In his study on this subject, Green (1978) asks, 'Who got what they wanted?' and argues that: 'The short answer to the question is probably nobody in Tanzania in full but quite a number of involved institutions in part. . . The results look more like a mildly unsatisfactory compromise than either a clever scheme to influence political decision taking, a cogent identification of and action to meet a compelling social priority, a production and investible surplus example to be copied or a major technical and financial debacle.'[7]

Conflicts of objectives also arise very commonly within a donor agency or among different offices of a donor government. Even where an appropriate choice of technology is recognised to be essential to the achievement of development objectives, development objectives often compete with commercial objectives, political objectives, and managerial objectives of the agency's administrators such as achievement of disbursement targets.

In a way, the life of an aid project may be then considered as a sequence of games which develop from one stage to the next through concessions of conflicting objectives, beginning with the stage of project design.

White's study also touches upon an important issue raised by Stewart, namely, the cumulative effect of technology choice. He quotes a study on tractors in Sri Lanka, which states that 'the process was reinforcing: once tractors were imported, the ownership, control and advocacy of them created new economic and institutional sources of support for the technological options they represented'. Referring to the Afghan case where a request for financing grain storage led to a series of proposals for the construction of an integrated network of transport and storage facilities, he argues further that 'while the influence of an agency on the first project may be slight, the impact of its subsequent decisions on how it follows up that initial project may be considerable'.

In the first part of his study, Watanabe attempts to assess the capability of indigenous technology generation currently existing in Third World countries, using as an index the number of inventions. This analysis casts some doubt on the feasibility of a strategy based on the build-up of a local capital goods industry, as proposed by Bruton. For in so far as inventive activity is captured in patent statistics, and to the extent that this activity is a *precondition* for the emergence of a capital goods sector, Watanabe's data are not very encouraging. Inventive activity is extremely limited in most Third World countries.

However, Watanabe's paper also suggests a way in which this problem may be overcome. He points out that the patent (including the utility model) system has been one of the main policy instruments that today's industrialised nations used for the development of indigenous technology capacity. In his view, the patent system is more promising as a means of encouraging people's participation in technology development, than the official R and D institutes which tend to remain academic ivory towers. The success of a patent system, however, depends not only on the mode of operation of the system, but also on a wide range of macroeconomic policies and programmes which create an adequate environment and incentives to people's inventive effort. His statistical analysis in the earlier part of the chapter suggests that basic education and policies pertaining to industrial growth may have special relevance in this regard.

Taken together, the papers by Bruton and Watanabe raise the important question of the appropriate *sequencing* of efforts to build up indigenous capacity. The emergence of such capacity, that is, may need to proceed through a quite well-defined series of stages, with improvements at one stage being to some extent contingent on the capabilities acquired at the previous stage. And policies that are appropriate at one stage may not, therefore, be suitable at another. Consider, for example, the stance that is to be adopted towards technology imports. Bruton regards technological dependency as a strategy alternative to the route aimed at the emergence of indigenous capacity. Watanabe, in contrast, argues that, 'Dependence upon imported technologies need not contradict the long-term goal of technological self-reliance.' Technology imports, in his view, enlarge learning opportunities. But the discrepancy between these views may be largely apparent, for which is appropriate will depend, among other things, on the particular level of technological capability that the country has already achieved.

EMERGING ISSUES

The individual authors in this volume raise a wide variety of issues that deserve the special attention of both policy-makers and future researchers. We will not attempt here to enumerate all these issues. Rather, we shall focus on only a few subjects that are either common to different chapters or which seem extremely important to us and yet are relatively neglected.

The first set of issues are related to the concept of indigenous technology capacity. Despite endorsement of the need for 'indigenous technological capacity building' by many authors, there is as yet no clear statement of what this really means at different stages of development and different sectors of an economy. Studies conducted elsewhere, notably those of the World Bank group led by Westphal and by the IADB/IDRC group led by Katz, have examined different aspects of technological capacity building in Third World countries, but their scope has been largely confined to the modern large sector. There is as yet no model for analysis of the sequential process of evolution of indigenous technology capacity in a given economy or industry. Nor are we yet clear as to what factors have a crucial influence on the development of such capacity at different stages of industrialisation. This is, however, vital for the purpose of spelling out concrete policy measures to be taken for building up the technological capacity of different countries. Among others, the following questions seem to warrant special investigation:

(1) What is the (quantifiable) definition of 'indigenous technological capability' at the macroeconomic level and at the sectoral level (e.g. with respect to the 'formal' and 'informal' sectors), which is the most relevant to policy objectives such as employment creation and poverty alleviation?

(2) What variables are crucial to the formation and expansion of such capacity?

(3) How such variables can be controlled through policy instruments and what factors tend to influence the effectiveness of such policy measures? (For example, even if R&D effort is identified as one of the important factors influencing the indigenous technological capacity, a given amount of R&D expenditure can create different impacts depending on how it is spent.)

(4) How and under what circumstances can shops for repair and maintenance be developed into proper capital goods producers; what are the crucial technological differences between the two, and how gaps between them can be reduced?

(5) What is the role of technology imports in local capacity building? What are the net social costs of pursuing a policy of protecting indigenous technologies in different sectors of the economy?

The second issue that we wish to emphasise concerns the potential for direct intervention by the government in the technology decisions of state-owned enterprises in developing countries. Compared to what Stewart terms indirect policies of intervention (such as those that operate through factor prices, exchange rates, etc.), direct policies have been neglected as an instrument of technology policy. Yet these policies would appear to have considerable superficial appeal. In many developing countries, public enterprises have come to dominate sizeable segments of the economies, and (at least in theory) the government ought to be able to exert considerable influence over these firms.

The experience of many developing (and developed) countries, however, has belied the simple assumption that the state encounters no obstacles in implementing its technology decisions because the public enterprise (and those within them) are agents who act merely as extensions of the interests of the state. In practice, it has proved extremely difficult to ensure that the behaviour of public enterprise responds to the goals towards which society wishes to move.[8]

There are two major issues for research arising out of this disappointing performance. The first concerns those developing countries in which, for political or economic reasons, publicly owned firms are held to be essential components of development strategy. In Tanzania, for instance, public enterprises are institutions that have a critical role to play in the transition to socialism. The relevant policy question for these types of countries is: how the technology decisions of such firms may be made to serve the goals of development more effectively. In other developing countries in which public enterprise is not desired for its own sake, one needs to be concerned instead with the conditions under which direct intervention represents a superior instrument of technology policy than more indirect measures (such as regulating the technology choices of privately owned firms). To answer this question, in turn, research could usefully attempt to ascertain the extent to which the problems that have been encountered with the direct form of intervention are *inherent* in the use of public enterprise as an instrument of policy, or whether they are more amenable to correction. The greater the extent to which the difficulties fall into the former category, the less appealing become the relative merits of policies for direct intervention.

Lastly, the question of choice among alternative policies or strategies, raised in James' chapter, seems to be of paramount importance for adequate development planning. An increasing degree of specialisation in the area of technology among development economists has led to an understandable tendency to pose the solution to policy problems too readily in technological terms. This tendency is not only unhelpful to policy-makers whose concern is (or ought to be) with the effectiveness of alternative instruments, but the indiscriminate advocacy of technology may even serve to undermine its credibility as a useful policy option. It seems to us, as crucial therefore, that future research should adopt a perspective that is considerably broader than much of the work that is currently being conducted. Just as one may often need to consider the relative merits of alternative technology policies in meeting particular objectives, so too should one raise the question of whether the objectives can be more effectively pursued by policies other than those that operate through the technology dimension. One aspect of this decision will clearly concern the political feasibility of alternative policies. In this respect applications of Stewart's costs–benefits matrix and Enos' game-theory approach to specific situations could prove highly valuable. But one also needs to take into account the administrative requirements associated with various policies and here too there is much scope for further research.[9]

NOTES

1. General surveys of the debate are contained in Morawetz (1974), White (1974), Jéquier (1976), Stewart (1977), Westphal (1978) and Bhalla (1981).
2. See, for example, Wells (1973), Morley and Smith (1977), Lecraw (1979), Pack (1982).
3. One exception is the study by Boon (1964).
4. See Green (1982), p. 90.
5. See Stewart (1981).
6. Research currently being undertaken as part of the World Employment Programme is concerned to illuminate the distinction between the technological behaviour of public and private enterprises through a set of structured comparisons in several developing countries.
7. See Green (1978), pp. 33–4.
8. See, for example, Perkins (1983).
9. See Uphoff and Ilchman (1972).

REFERENCES

Bhalla, A.S. (ed.) (1981) *Technology and Employment in Industry*, 2nd ed (Geneva: ILO).

Boon, G.K. (1964) *Economic Choice of Human and Physical Factors in Production* (Amsterdam: North Holland).

Green, Reginald Herbold (1978) 'The Automated Bakery: A Study of Decision Taking Goals, Processes and Problems in Tanzania', Working Paper No. 141 (Brighton, Institute of Development Studies, University of Sussex), October.

Green, Reginald Herbold (1982) 'Industrialisation in Tanzania', in Martin Fransman (ed.) *Industry and Accumulation in Africa* (London: Heinemann).

Jéquier, Nicolas (ed.) (1976) *Appropriate Technology, Problems and Promises* (Paris: OECD Development Centre).

Lecraw, D. (1979) 'Choice of Technology in Low-Wage Countries: A Non-Neoclassical Approach', *Quarterly Journal of Economics*, November.

Morawetz, David (1974) 'Employment Implications of Industrialisation in Developing Countries: A Survey', *Economic Journal*, September.

Morley, S.A. and Smith, G.W. (1977) 'The Choice of Technology: Multinational Firms in Brazil', *Economic Development and Cultural Change*, January.

Pack, H. (1982) 'Aggregate Implications of Factor Substitution in Industrial Processes', *Journal of Development Economics*, August.

Perkins, F.S. (1983) 'Technology Choice, Industrialisation and Development Experiences in Tanzania', *Journal of Development Studies*, January.

Stewart, F. (1977) *Technology and Underdevelopment* (London: Macmillan).

Stewart, F. (1981) 'Arguments for the Generation of Technology by Less-Developed Countries', in *Annals* of the American Academy of Political and Social Science, November.

Wells, L. (1973) 'Economic Man and Engineering Man: Choice of Technology in a Low Wage Country', *Public Policy*, Summer.

Westphal, Larry E. (1978) 'Research on Appropriate Technology', *Industry and Development*, No. 2.

White, Lawrence J. (ed.) (1974) *Technology, Employment and Development* (Manila: The Council for Asian Manpower Studies (CAMS)).

Uphoff, N. and Ilchman, W. (eds) (1972) *The Political Economy of Development* (Berkeley: California University Press).

Part I
Conceptual Approaches

Part 1
Conceptual Approaches

2 Macro Policies for Appropriate Technology: An Introductory Classification

FRANCES STEWART[1]

INTRODUCTION

In every case, an actual decision about technology takes place at the micro level. That is to say, when an investment decision is made, which embodies a particular technology, this decision is made by decision-makers within a productive unit. The productive unit in question may be a large organisation with worldwide activities, such as a multinational corporation, or it may be a very small unit, such as a family firm or farm. Whatever the size of the unit, the level of decision-making is defined here as *micro*, since the decisions are taken by the units in question in the light of their own objectives and resources.[2] But while each decision takes place at the micro level, these decisions are strongly influenced by the external environment in which they take place. Apart from direct interventions – when governments themselves make particular investment/technology decisions – governments can only influence technology decisions by influencing this external environment. Direct interventions tend to be confined to a small part of total technology decisions, especially in a mixed economy. Hence government's greatest potential influence on technology decisions takes the form of influencing the decision-making unit's external environment. Despite this, the appropriate technology institutions have tended to confine their activities to direct intervention.[3] In contrast, this paper considers government policies to affect the environment in which micro decision-making

19

takes place, with a view to classifying/identifying the set of policies likely to promote appropriate technology. These policies are described as macro policies to emphasise their general nature, as distinct from direct and particular intervention at the micro level.

DEFINITIONS

It is easy to spend too much time on definitions, particularly as often we know what we are talking about even where precise definitions are impossible. Nonetheless, it is necessary to be as clear as possible about the two main concepts under discussion – *macro policies* and *appropriate technology*.

Macro Policies

Macro policies, as used here, cover all those general government policies which influence the environment in which micro decision-making units operate. The micro decision-making units in question are normally described as 'firms' in microeconomic textbooks. In this paper the rather clumsy term *micro decision-making units* is preferred because the units in question do not consist solely of conventional private sector firms, with owners, employers and employees, but also include public sector firms, co-operatives, and family and household organisations. For brevity the words 'units' and 'firms' will also be used to encompass this spectrum of micro decision-making units.

The macro policies at issue then extend far beyond the normal macro policies of economic textbooks, which are largely confined to policies towards major economic aggregates, such as the money supply, interest rates, public expenditure levels and budget deficits. While these policies are one subset of the macro policies considered here, we also include many others – such as policies towards technology supply, market access, and so on. *Macro policies* are used to describe this great variety of policies to indicate that they apply across the board to the whole spectrum of micro decision-making, in contrast to *particular* interventions. The policies in question then might equally well be described as general policies towards appropriate technology. In what follows the two terms, *macro policies* and *general policies* will be used interchangeably.

Appropriate Technology (AT)

A very great deal has been written on the question of the definition of *appropriate technology*.[4] We may contrast two views on its definition. On the one hand, there are those who define appropriate technology with reference to welfare economics. According to this view, appropriate technology is that technology which maximises social welfare. 'Appropriate Technology may be defined as the set of techniques which make optimum use of available resources in a given environment. For each process or project, it is the technology which maximises social welfare if factor prices are shadow priced.'[5]

In contrast, most of the AT groups identify AT with a specific set of characteristics, rather than the general phenomenon of social welfare maximisation. This is also the approach adopted by some social scientists.[6] The precise list of characteristics included varies among authors. Among *economic* characteristics, a more appropriate technology is normally defined as being

- more labour-using than a less appropriate technology (higher L/O);
- less capital-using (lower K/L);
- less skill-using;
- making more use of local materials/resources;
- being smaller scale;
- producing a *more appropriate product* (i.e., a simpler product designed for lower income consumers, or a product suitable as an input into other appropriate technology).

Other (less economic) characteristics that are sometimes emphasised are that appropriate technologies should not be environmentally damaging (e.g., Farvar (1976)) and that they should fit in with socio-economic structures of rural life (Reddy (1979)).[7] Because of the multidimensional nature of the 'specific characteristics' definition, according to this approach technologies may be appropriate in some respects, inappropriate in others. Moreover, since societies differ in material resources, culture and socio-economic structures, technologies may be appropriate for some societies while inappropriate for others. The specific characteristics approach thus does not uniquely, and for all countries and for all time, identify particular technologies as appropriate or inappropriate; rather it points to a multidimensional set of characteristics which tend to be associated with more appropriate technologies.

Both definitions of appropriate technology – the social-welfare definition and the specific characteristics definition – have their disadvantages. On the one hand, the specific characteristics definition could lead to a situation in which the appropriate technology (i.e., the one with a more appropriate set of characteristics) is in fact an inferior technology and/or one which does *not* best enable a country to meet its objectives. This might arise because the technology with appropriate characteristics was very inefficient in a technical sense (of low productivity) compared with a technique with inappropriate characteristics. In some cases, the technique with appropriate characteristics might still be the best choice (despite technical inferiority), if its effects on some objectives (e.g., enabling the very poor to participate in economic activity, or in terms of beneficial environmental effects) outweighed its low productivity. But in others this might not be the case, and therefore the technology with inappropriate characteristics should be preferred to the appropriate technology. The specific characteristics approach also encourages concentration on the immediate consequences of a technology, while neglecting the wider effects which might make a technology with inappropriate characteristics preferable. For example, a capital-intensive technology might generate exports and thus relax a foreign exchange bottleneck, permitting more employment in the long run than a more labour-intensive technique.

In such situations, use of the specific characteristics definition may lead to one of two outcomes: *either*, it would encourage the adoption of the 'wrong' technology, taking all aspects into account; *or*, it would mean that techniques with appropriate characteristics should not always be selected – i.e. the appropriate technique is not always the best choice. Apart from the fact that this involves a sort of linguistic contradiction, it also means that additional criteria (of an efficiency, social–cost–benefit type) are necessary in order to decide what the best choice of technique is.

The social-welfare definition avoids these problems: the appropriate technology is always (by definition) the right technology to choose. But there are also difficulties with the social-welfare definition: first, it requires the whole apparatus of social welfare/cost–benefit analysis to determine what is appropriate technology. This apparatus involves many conceptual and practical problems, especially with respect to determining social values and shadow prices.[8] Secondly, it can conceal, or even eliminate, the insights offered by the appropriate technology movement, in the economists' labels of 'social welfare', 'objective functions', which, while all embracing, also tend to be contentless, lacking descriptive connotation. Thirdly, it does not contain any 'signposts for action' in the way that the specific characteristics definition does. Thus, using a social-welfare definition, one technique out of any set would maximise social welfare and is therefore appropriate. Such a technology could be large scale, capital intensive and ecologically damaging. None of this would be indicated by the definition. But indicating that *all* the available techniques are inappropriate in some respects is extremely important in demonstrating the need for search and research in specific directions and specific areas. This 'signposting' is one of the major contributions of the appropriate technology movement and one that is lost by adopting the social-welfare definition.

The discussion of the relative merits of the two approaches has implications for methodology, whichever definition is finally adopted. On the one hand, if a specific characteristics approach were adopted, then other criteria (encompassing something of the social-welfare approach, albeit rather simply) would be necessary to make sure that social and economic consideration of efficiency, dynamic as well as static implications, and general as well as partial effects were included in determining technology choice. On the other hand, if a social-welfare definition were adopted, then it would be essential to supplement this with a set of specific characteristics, and an assessment of each technique in terms of this set, in order to provide signposts for choice, search and research.

Both approaches have to be applied in a *dynamic* context. Over time countries' resource availability, as well as technological possibilities, are changing. Consequently the appropriate technology (under either definition) will generally change over time. Moreover, the time dimension also enters in terms of the effects of technology choice on a country's development path over time. This arises most obviously in connection with employment creation. The technique which maximises short-run employment may be different from that which would be most likely to maximise long-run growth in productive employment opportunities. Which is the preferred technique then depends on the trade-offs involved and also country preferences in relation to time.

The question of preferences/objectives is at the heart of a social-welfare approach to appropriate technology since social welfare can only be measured once social objectives have been determined. Country objectives do not enter explicitly into the specific characteristics approach. However, in the selection of characteristics a certain set (although not precise details) of objectives has been assumed, namely, a spreading of employment opportunities, the provision of income earning opportunities to the poor and consequent reduction in poverty and inequality. This preselection of objectives can be regarded as a strength of the approach, because it truly reflects a major element in the appropriate technology movement, which is concerned with objectives as much as the means of achieving them. Yet it pre-empts what should be the choice of the people concerned rather than that of an outside movement. Moreover, while the broad elements of the objectives embodied in AT would be widely agreed upon, the precise details are by no means settled, and especially the weight to be attributed to different objectives. This lack of clarity is reflected in the specific characteristics approach, in the ambiguity about precisely which objectives to include, and the weight to give the various characteristics, where conflicts occur. With a specific characteristics approach, a government may reject the appropriate technology because it does not share AT objectives – for example, because it wishes to modernise and maximise growth and places little weight on reducing poverty. In contrast, with a social-welfare approach, government objectives are automatically incorporated in the measure of social welfare. This means a government would never reject the AT, as identified by the social-welfare approach, but the AT could be quite unrecognisable in terms of the normal AT objectives – for instance, it could involve heavy expenditure on prestige and capital-intensive modernisation projects if that is what the government wants.

In the rest of this paper, a specific characteristics approach is adopted, while recognising (a) that the selection of precise characteristics, and their weighting, is often arbitrary and should always incorporate local preferences; (b) that the characteristics need to be supplemented by some assessment of social efficiency before actual choice of technique is made; and (c) that it is perfectly possible that the AT, as defined with this approach, could be rejected because implicit objectives involved in the characteristics chosen do not reflect the actual objectives of a particular government.[9]

THE MACRO ENVIRONMENT AND THE MICRO DECISION

We define the micro unit as being the organisation (generally a firm) which actually makes a technology decision: i.e., chooses, acquires, installs and operates a particular technology. This unit may be in the public or the private sector, may be capitalist or a family unit, in industry, services or agriculture. The unit makes its decision in the light of the following factors:

(1) Its own objectives (which may be to maximise pre- or post-tax profits, to maximise family income, to maximise employment, to satisfice . . .).

(2) The resources available to it, and the prices it has to pay for these resources. The resources include material inputs, labour of various skills and capital equipment. The prices an organisation has to pay consist of the actual prices it faces (e.g., wage rates for labour) and also taxes of various kinds (e.g., social security payments for labour).

(3) The nature of the market it faces, whose major dimensions are size, industry and type. By the *type* of market is meant the sort of market that makes up the major consuming element – high-income or low-income, for example, local or international. In part, the nature of the market is dependent on the history of the organisation. Thus a firm that has always manufactured soap tends to stay in that market. A large-scale firm that exports most of its products will be concerned with the international market, and so on. But the market is also a variable, which may be changed by activities of the firm.

(4) The organisation's knowledge of available technological alternatives. This is a function of three variables: (i) the actual 'technology shelf' in the industry, which depends on the historical development of technology in that industry; (ii) information channels in being about technology in that industry (e.g., consultants, machinery salesmen, information services, etc.); and (iii) the efforts the firm puts into collecting information and the nature of those efforts. For example, one firm may do nothing about acquiring information and simply depend passively on local machinery salesmen for knowledge; others may actively search, sending engineers abroad to seek suitable second-hand machinery, or using their own research and development to produce new knowledge.

We may summarise by saying that technology choice is a function of firm objectives, resource availability and cost, markets and technology (or O,R,M,T). It follows that macro policies may affect micro choice of any given unit by affecting any of these four variables. In addition, the government may affect the balance of choice in the economy as a whole by altering the proportion of resources controlled by various units, or what we shall call the *composition of units* in the economy. For example, it is established that small-scale units generally use more labour-intensive technologies than large-scale, whether they are public or private. By shifting the control of total resources the government may bring about more appropriate technology in the economy as a whole, even though it has not altered the actual decisions of any particular unit.

Macro policies, considered briefly below, then consist in the policies which affect the four variables and those affecting the composition of different units in the economy.

POLITICAL ECONOMY

The socio-economic framework of an economy is highly relevant in two respects. First, the framework influences many of the variables affecting technical choice. For example, the organisation of productive units, their access to resources and the nature of the markets they serve are heavily influenced by the system of political economy. In most contexts, the broad dimensions of this framework have to be taken as a given, not a variable. Consequently, conclusions as to appropriate macro policies may vary according to the socio-economic framework. Secondly, political economy variables influence which policies are feasible in a particular context, and which are not. Thus a strategy which might appear first best from the point of view of promoting appropriate technology may in fact have to be ruled out in a particular context because it appears infeasible, and some apparently second best set of policies be recommended. This is a problem-ridden area because it is very difficult to judge what is and what is not feasible. Yet to ignore it can lead to useless recommendations.

Two somewhat extreme views, with respect to political economy, can be contrasted. On the one hand, economists' first-best recommendations normally completely ignore or assume absent limitations/constraints imposed by political economy. In many situations, to assume no constraints is likely to mean that recommendations are ignored. Rudra (1982) indicates the failures of appropriate technology in

India since independence – despite government rhetoric in its favour – and attributes such failures mainly to conflicts with important power groups. A similar dichotomy between rhetoric and action can be seen in Tanzania and the explanation is also largely a matter of political economy.[10] In contrast to the 'no constraints view', the view is sometimes taken that the forces of political economy are such as to leave very little room for manoeuvre.

Galtung's view of *structures* comes close to this position (Galtung, 1980).[11] Every technology, according to Galtung, is associated with a certain structure, economic, social and cognitive. The structure associated with a technology, 'produces, filters out and accepts only the techniques that will be accompanied by such structures, thereby reinforcing the structures themselves'.[12] The structure associated with Western advanced technology consists of the capitalists, the scientists and the bureaucrats, who benefit from the technology and whose positions would be threatened if an alternative technology were adopted. But these groups are precisely those who make most of the technology choices: consequently, there is very little possibility of introducing an alternative appropriate technology on any scale, in any society where such advanced technology is well established.[13]

Historic experience – and common sense – suggests that there is a good deal of truth in a structures or political economy view that leaves rather little room for manoeuvre in terms of choosing alternative technologies. But it does seem that there is some possibility of choice – as indicated by the wide variety of technologies observed in practice in use, both within a single country and across countries.[14] Taking these forces into consideration suggests that more effort needs to be devoted to exploring the area (and determinants of freedom of choice). One fruitful approach is a game-theory approach, as suggested by Enos in this volume.

It is first necessary to identify and delineate the groups affected by alternative technology choices, and also to determine the *power* of the various groups over the technology choices. No group is likely to be completely homogeneous. For example, workers are differently affected according to their skills, their experience and versatility, their family circumstances, and so on. Employers, similarly, will be differently affected according to their particular circumstances. Nonetheless, it is usually possible to pick out the main groups which are likely to be affected in particular ways by technology choices, even though there may be differences within the groups, and individuals who belong to more than one group.

We may distinguish between decisions involving the government and those that do not. Decisions that do involve the government are those where the government itself is playing an active, *direct* role (e.g., in choice of technique with respect to public enterprises, or, more directly, public expenditure) and those where the government is playing an *indirect* role by forming, or at least influencing, the rules (prices, tariffs, etc.) which determine the framework in which all micro units take decisions. It is this last category which is the relevant one for the purpose of the formulation of macro policies for appropriate technology, although the considerations are similar in cases where the government is acting directly. In all such decisions, obviously one of the relevant groups is the government. The question at issue then is the nature of the government in terms of the interests it represents and hence *how* it is affected by different decisions and also its *objectives* and the *game* it is playing, which together will determine the actions it takes.

Implicitly, the two extreme views identified above are making very different assumptions about the nature of government. The first-best type of view of many economists, implicit in much social–cost–benefit analysis, is that the government is a homogeneous entity whose interests are those of the nation and whose objectives are to maximise social welfare. This leaves (an impossible)[15] task of identifying the interests of 'the nation', especially in contexts where (to use Disraeli's term) there are 'two nations', in order to make maximisation of social welfare meaningful. If these problems are overcome completely then the approach avoids the need to trace the particular interests that are embodied in particular governments and the effects these have on technology choice and associated policy variables. At the other extreme, we have the view that the government unambiguously (and invariably) represents *one* set of interests – in Galtung's example, those of the researchers, bureaucrats and capitalists. But this view too seems an oversimplification of a complex world in which governments are not homogeneous and have ties with different groups in different societies; not only do the particular ties vary but the tautness of these links also vary, particularly where conflicting interests occur among the various elements which influence government. Careful empirical work is needed in each situation to delineate the various groups which make up/influence government decision-making. Such research is a prior requirement before either of the two extreme views just described can be accepted or rejected, despite the fact that exponents of both views do not seem to view such research as necessary.

If, as seems likely, governments are linked to many different groups, then there is potential for conflict in interests among them. In looking at potential rule changes (or macro policies), and identifying the potential gains/losses for various groups, three types of rule change will emerge:

(1) Those where all groups constituting government interests gain;
(2) Those where all groups would lose;
(3) Those where some would gain, some lose.

It is helpful, in exploring feasible strategies, to classify desired changes into these three categories. In so far as changes can be identified which positively affect *all* major parties (category (1)) then these are obviously a politically realistic set of changes. It might be thought that very few changes would actually fall into this category, since any such changes would presumably already have been put into effect. But in a dynamic world, where new possibilities are continuously arising (which is especially the case for technology), new information may lead to the identification of such changes. It is useful to identify the second category (those where all lose) in order to put them on one side as not politically feasible. A large number of changes are likely to fall into the third category – some gainers, some losers. It is here where most of the potentially feasible changes are likely to lie. The size of this category and the possibility of change within it will determine the extent of politically feasible macro policies for appropriate technology in any particular context. Two facts prejudice the likelihood of change within this category. First, the very fact that there are losers indicates that the changes will not be politically straightforward. Secondly, the fact that the changes have not been carried out may suggest that the losing interests dominate over the gaining. Nonetheless, it is to this category that changes in macro policy need primarily to be directed. Once the category has been identified, it may be possible also to identify actions which may make the changes more likely. These include (a) altering the proposed reforms so that the distribution and size of gains and losses alter, in order to make the changes more acceptable; (b) devising new political coalitions to strengthen the gainers *vis-à-vis* the losers; and (c) providing information which may alter the various parties' assessment of the changes.

A simple example may make the considerations discussed above clearer. Assume that tractors are being introduced in large farms in a particular area, and that these tractors are being bought with subsidised credit available to the large farmers from government credit institutions. In terms of the specific characteristics definition of appropriate technology, it appears that the tractors are inappropriate compared with the use of labour plus bullocks, since the tractors are more capital intensive and large scale, use imported inputs, etc. Are the tractors also inappropriate with respect to the alternative 'social welfare' definition? This depends on the values (shadow prices) of a social welfare exercise but in many cases,[16] it has been established, tractors do *not* maximise social welfare. Let us assume that this is the case here. Hence, to ensure the AT is selected the government needs to alter various rules so that tractors are no longer introduced. The change in rules could consist in

(a) changing land distribution with land taken from the large-scale farmers who consequently are no longer relevant to technology choice, while the small-scale farmers who acquire the land continue to make appropriate decisions;

(b) changing the availability of credit to large- and small-scale farmers;

(c) changing the price of credit and/or the price of labour so that the large farmers find it more profitable *not* to buy tractors;

(d) banning the import of tractors.

All these, (a)–(d), have been suggested at various times as a means of promoting appropriate technology. The choice between the various alternatives can be approached at various levels and from various points of view. One approach is the *economists'* of maximising efficiency. In general, this would tend to favour (c), in so far as it could be shown that previously the prices were distorted. Alternatively, those primarily concerned with *distributional* considerations would favour (a) as a first best (redistribution of land), followed by (b), redistribution of credit. While not totally disregarding considerations of economic efficiency or distribution, a *political economy* approach would look at (a)–(d) rather differently from the point of view of feasibility. Using the approach suggested above, the first requirement would be to draw up a matrix showing the gains and losses to various groups from the various changes. This has been done in Table 2.1 (for illustrative purposes only).

In the matrix category (d) has been divided into two: (d)' banning *all* tractors, including local production; and (d)" banning imports. While obviously banning all tractors is more effective from an AT point of view, practice (d)" – banning imports – is the more common

TABLE 2.1 *Costs and benefits matrix*

Groups affected	Rule changes	(a) Land redistr.	(b) Credit redistr.	(c) Price changes	(d)' Banning tractors	(d)" Banning tractor imports
Large land owners		− 100	− 20	− 10	− 5	− 2
Small land owners		+ 50	+ 20	+ 5	0	0
Landless labourers		+ 70	+ 20	+ 20	+ 20	+ 5
Tractor producers (foreign and local)		− 30	− 30	− 30	− 50	− 50* + 40**

* Foreign tractor producers
** Local tractor producers

policy. The effects on tractor producers then is divided where relevant, into effects on foreign producers and on local producers.

The matrix illustrates a conflict situation, with some groups gaining, some losing from each change. As is pointed out, the size and distribution of gains and losses among different groups differ markedly according to the strategy selected. The redistributive strategies (a) and (b) hurt the large landowners most and help the small landowners most. The landless labourers gain most from the redistribution of land strategy because it is assumed that some of the land is redistributed to them. They benefit equally from each of the other strategies except (d)" because banning tractor imports is less beneficial to them as it is assumed that some tractorisation continues with locally produced tractors. Nothing is said in the matrix about the *social* benefits of the various alternatives because any adding up exercise depends critically on the weight attributed to various groups.

In order to say something about the political feasibility of the various strategies we need to know the weight the government gives to each interest group. This requires an analysis of the political composition of the government, the sources of its support and the power of pressure groups. Governments whose major concern and support is from landless labourers would choose (a); in contrast governments which are dominated by large-scale farmers would do nothing. In practice most governments are subject to a variety of interests of varying strengths; they are concerned about employment and therefore give some importance to landless labourers, but they are also subject to (large) farmers' pressure and therefore would tend to choose options which minimise the cost to them (i.e., (c) or (d)); in addition many governments are subject to pressure from local (tractor) producers. Consequently (d)" may prove the most attractive option – since it benefits a major local interest group (tractor producers) while harming large farmers only by a small amount and giving the appearance (and perhaps a little the reality) of helping the employment situation.

It is clear then that what is politically feasible cannot be assessed independently of a careful empirical analysis of an actual situation: this analysis needs to include an assessment of the distribution and extent of gains and losses to various groups and also the nature of government support. In many situations, however, it appears that there can be conflicts between first-best options from the point of view of economic efficiency and/or social justice (options (a) and (c) in our analysis) and those which are politically feasible (option (d)"), in at least one plausible political scenario. Where such conflicts emerge – which can prevent or dilute action to promote AT – it is worth investigating whether changes can be introduced which might alter the outcome. This might be achieved by altering the matrix of benefits and costs or by altering the way the political system processes this matrix. For example, on the one hand a new technology might be produced which would benefit *all* categories: one example might be the manually propelled tractor which could be made locally, benefiting local tractor producers and need not be to the disadvantage of large landowners, if associated with some change in taxes/subsidies to make it attractive. On the other hand, political action on the part of the landless labourers, perhaps combining with small land-owners, might alter the weight the government places on costs and benefits of various groups and therefore the outcome with a given benefit matrix. Possible and effective changes will vary according to the particular case.

It must be emphasised that the example given above has been illustrative. It is close enough to actual cases, however, to help explain why tractorisation (and other types of inappropriate technology) persists despite evidence of negative social benefits. The example is intended to emphasise the conclusion already reached: that designing policies independently of actual political realities may often lead to ineffective recommendations.

The area of political economy impinges on the timing and sequencing of particular strategies. Certain developments (e.g., technical choices) establish a nexus of interests which then become powerful in influencing future decisions – both from a technical and a political point of view. For example, the establishment of an assembly plant may (i) establish employer and employee interests in continued protection of that industry and its market; and (ii) create technical needs and political support for ancillary services (e.g. transport, local maintenance service, etc.). In the above example, an initial decision to establish a local tractor industry sets in force political forces which alter the likely choice of strategy. The initial decision should then take into account likely subsequent developments. A different initial strategy may become optimal if subsequent developments are allowed for. This conclusion seems widely applicable in the area of technology, where there are strong links (both political, technical and economic) between choice of technique at different stages of production. Decisions about choice of technique at one stage then provide strong justification for choosing compatible techniques at earlier or subsequent stages.

CUMULATIVE FORCES

Political and technical forces tend to lead to cumulative decisions with respect to each of the four variables delineated above, O,R,T,M. To summarise briefly, O is mainly a function of the nature of the micro unit, and of the competitive structure of the economy. The existence of any particular set of organisations (e.g., foreign-owned firms, family firms) establishes political pressures for their continued existence; similarly, a highly protected non-competitive structure of the economy generates interest groups whose survival depends on the continuation of this policy. Conversely, a strongly competitive structure forces firms increasingly to become profit maximisers, rather than satisficers.

R represents access/price of resources; once some sectors receive favourable access (e.g., subsidised capital), this becomes a difficult policy to reverse. Similarly, a 'high'-wage policy is difficult to reverse.

M, market access tends to be cumulative in the sense that it is much easier to maintain than create a particular market; consumer tastes may be (irreversibly) created for a particular set of products.[17] In addition, firms acquire experience in serving a particular market.

T, which is knowledge about techniques, is strongly cumulative. Research, development and knowledge about techniques tend to be located around techniques in use (or already known about) so that the direction of technical change is heavily influenced by techniques in use.

The cumulative forces described here mean (i) that it may be easier to change strategy early in the development process before too many cumulative forces have been established; and (ii) that marginal change may be difficult to achieve. A radical break has to occur to generate cumulative forces operating in a different direction.

CLASSIFICATION OF MACRO POLICIES

The description of variables influencing technical choice at the microlevel serves as a way of identifying and classifying macro policies. In general such policies have the twin objectives of altering the decisions of particular units, and of altering the composition of units in the economy towards those units likely to choose more appropriate technologies.

Objectives

The objectives of any micro unit depend on three things. First, on the mode of production the unit represents. Objectives differ for example, as between foreign-owned firms (aiming to maximise worldwide profits) and local firms (concerned with local profits); as between family firms (aiming to maximise family income) and capitalist firms (aiming to maximise profits); as between publicly owned and privately owned firms, and so on. Second, objectives may differ according to organisation within the firm; in some firms 'engineering' man may predominate, in others 'economic' man (to use Wells' (1973) terms), while in some types of firm managerial objectives such as output maximisation have come to dominate over capitalist objectives of profit maximising. Third, firms' objectives tend to vary with the economic environment in which they operate because in some environments some objectives may not permit the firm to survive. For instance, in a protected and oligopolistic environment firms may be able to pursue 'satisficing' objectives, while in a more competitive environment they may be forced to be profit maximisers to survive.

Government policy can influence objectives in a number of ways: directly, via government directives to publicly owned enterprises; and indirectly, by changing the economic environment; and for the economy as a whole, by changing the composition of units.

It is not at once obvious which objectives best serve AT – indeed, it is apparent that the effect of the objectives pursued depends on the interaction with the other variables, so that in some contexts profit maximising could lead to appropriate choices, while in other contexts it may result in inappropriate choices. In addition, short- and long-run consequences may differ: in the short run if firms in the public sector aim to maximise employment, in this dimension they will select appropriate technologies, but if this involves the use of highly inefficient methods it could damage the long-run possibilities for appropriate technology. It is nonetheless possible to draw a few conclusions on objectives:

(a) 'Satisficing' type objectives tend – especially in the case of large firms reliant on foreign technology – to lead to inappropriate choices. According to some research (e.g. Wells, 1973) 'engineering' man is thereby empowered to make the effective decisions, which tend to be for over-sophisticated technologies. According to others (Morley and Smith, 1974), foreign-owned firms tend to select technologies similar to those adopted in the home (developed) country. The choices of firms which are profit maximisers will depend on the external environment (prices, markets, etc.) they face. In some environments (oligopolistic competition with product differentiation, high wages, subsidised capital), they will tend to make inappropriate choices; in others they may make more appropriate choices. Hence policies towards changing objectives need to be accompanied by appropriate policies towards the environment.

(b) Foreign-owned firms (or ones with close foreign associations) are generally concerned to maximise worldwide (after tax) profits. This frequently involves adoption of the technology developed by the firm in its home country. Where the country in question is a developed country, this often means inappropriate technology and products. Where the country is another developing country, the technology may be more appropriate. Government policies here then may consist in (i) countering such effects by incentives/disincentives to encourage use of local technology; (ii) encouraging LDC multinationals; and (iii) shifting away from foreign ownership.

(c) In principle, governments may direct publicly owned firms to take particular decisions, but this is much too cumbersome if it involves daily interventions. Alternatively, and more efficiently, governments can lay down the criteria public firms should adopt. While this does occur in principle, in practice it is often ineffective because of the day-to-day pressures on public firms and within them which prevent them carrying out government policy. Thus even in countries whose governments are explicitly in favour of appropriate technology, public firms very often take inappropriate decisions. Public sector firms seem to be *more* rather than less subject to internal and external pressures which prevent them taking either appropriate or profit-maximising decisions.[18] This seems an area where additional work is needed. We need to know: (i) what the appropriate criteria of choice should be; and (ii) given the political economy of the public sector, what the best way of achieving this would be.

(d) Small-scale and family sector: this sector in general appears to make more appropriate choices in each dimension. One reason is objectives, but other factors (markets, access to resources) also play a significant part. Macro policies then need to be designed to promote this sector, and to increase its efficiency. It is not then a question of changing objectives, but of other policies (to be considered below under the other categories).

(e) Co-ops. etc: The impact of co-operative arrangements on technical choice and income distribution depends critically on the form of the co-operative – its objectives and resources. Empirically a range may be observed from those that consist in an association of members which have virtually no effect on resource allocation (e.g., marketing co-ops in parts of Africa) to the more ambitious units which affect objectives and resource allocation radically (e.g., Kibbutz, Communes).[19] For AT, government policy needs to promote a structure of co-operative arrangements that is likely to lead to appropriate choice.

Resources

This encompasses policies towards prices of factors and factor supplies. The price package associated with AT is well known.[20] It consists in eliminating the subsidies on capital (through tax allowances, and artificially low interest rates to the modern sector); and in trying to eliminate those factors which lead to high wage costs to the modern sector, including the various social security costs as well as the wages themselves. Exchange rate policies are also relevant;

overvalued exchange rates accompanied by high protection tends to reduce the price of imported capital goods.

It should be noted that these policies – basically raising capital costs and reducing wage costs – are all directed at the modern sector. For the rest of the economy, policies with the opposite effects are required. The informal sector and the traditional sector in general require lower interest rates and more access to capital. A major element in policy in changing the composition of units towards decision-makers which make more appropriate technology choices is the distribution of credit: through credit distribution, investible resources can be directed towards the small-scale sector and away from the large-scale sector.

Other aspects of resource policy are those related to the supply of resources used in appropriate technologies. The 'human capital' element is important here including policies towards training to generate appropriate skills. In some industries, supply problems with the skilled labour requirements of more appropriate technology are a significant reason why more capital-intensive technologies are introduced.[21] As far as capital is concerned, the promotion of a capital goods industry producing appropriate techniques can be of significance especially in areas where the international supply of appropriate capital goods is deficient, or where local adaption to international designs is necessary.[22]

Markets

Markets are important with respect to *scale* and to *type*. As far as scale is concerned, larger-scale production is almost invariably associated with more capital-intensive techniques.[23] Decisions on scale depend on resource access as well as markets. But markets do play an independent role. Where markets are localised (e.g., because transport costs are high or for other reasons) then small-scale and labour-intensive techniques are often selected. Encouragement is needed for rural industry, as against leaving the rural sector dependent on the urban centres for supplies.

The viability and dynamism of (dispersed) rural industry depends in part on the dynamism of agriculture. Where agricultural incomes are increasing fast, then local markets will also be rising. But while this may be a necessary condition, it is not sufficient to ensure that these markets are supplied by local small-scale sources. For this the development of local infrastructure (financial and technical services, marketing and marketing services) is required. In one study, the success of rural industry has been attributed in large part to infrastructural and organisational investments in the rural areas. In 1971 non-agricultural income accounted for over half of rural income in one area of East Asia, compared with between 15 and 20 per cent in the Philippines, where the bias against agriculture had been greater and investment in rural infrastructure substantially less (See Ranis, 1983).

Transport policy is also relevant: namely, the structure of roads (trunk roads or localised feeder roads) and the way in which transport charges are calculated, which in many cases encourages long haulage even where it is uneconomical. In India, for example, the system of pricing on the railways is such as to give substantial advantage to large-scale cement production by eliminating the real advantage – in terms of transport costs – that small-scale plants should have in their locality.[24]

As for *type* of market, here it is a question of the nature of the products in demand. The question of the nature of the products is itself a very important direct aspect of AT, with 'appropriate products' being one defining characteristic of AT. Appropriate products are products whose characteristics are such that they are broadly in line with the income levels and needs of the majority of the population. Products embodying very high income characteristics – such that only a small minority in a society will be able to enjoy them for a considerable time – are defined as 'inappropriate products'. This is not a watertight definition, of course, but nonetheless gives some broad criterion for assessing the appropriateness of products. Economies produce products not only for immediate consumption but also as intermediate products (inputs into further production) and also for export. The criterion of appropriateness just described does not apply to these categories. In the first category – intermediate products – appropriateness depends on whether the product is an input into appropriate or inappropriate techniques. For the second category, the relevant questions relate to the production techniques associated with the products, rather than the characteristics of the products themselves. All products, whatever their purpose, are associated with a certain (normally limited) range of production techniques; in general, once the characteristics of a product have been determined with detailed specification there is not much choice of production technique. Hence – quite apart from the appropriateness or inappropriateness of the products – markets are important in helping determine production techniques.

Naturally, upper income markets tend to be associated with products with international brand names. Within any product class then, these products are often (not always) more capital intensive. In addition, there is some tendency for high-income markets to consume products from more capital-intensive product groups.

Redistributive policies, raising the purchasing power of low-income groups and reducing that of high-income groups would tend to increase the demand for appropriate products and also to tilt production techniques towards more labour-intensive methods. In some cases, however, high-income groups consume more labour-intensive products, as for example with craft products such as hand-made carpets, hand-loom textiles. There are also direct policies towards markets/products, such as banning certain inappropriate products (e.g. powdered baby milk), limiting advertising, or promoting products with more appropriate characteristics/techniques.[25]

The international market tends to require a higher standard and more capital-intensive product within any product market, but this may be offset by the selection of more labour-intensive groups (e.g., textiles) as a consequence of international trade. Trade between developing countries may consist in more appropriate products (and production techniques) than trade between North and South, but again while this is probable within any product group, it may not apply to trade in manufactures as a whole. International trade policies then are relevant to the determination of markets, in both quantity and type, but their effects are complex and will depend on the precise policies and circumstances.[26]

Different organisations tend to serve – and even create – different markets. Although each type of organisation sometimes does serve each type of market, there is a tendency for the following links between organisation and market:

Foreign-owned firms and joint ventures	international markets	middle- and upper-income markets locally
Public sector		middle- and upper-income market locally (some international)
Large-scale private sector		
Informal sector Traditional sector		local low-income market

There is some potential for changing the markets which each type of organisation serves; there is also potential for changing markets by changing income distribution and the international trade environment.

Technology

This includes knowledge about the available technological choice. A major area for policy consists in promoting 'appropriate' information channels. The present system of information tends to be systematically biased towards recently developed advanced-country technologies (through salesmen, journals, etc.). Old appropriate technologies are rarely promoted; information channels between LDCs are also often weak. Informational channels need improving nationally and internationally. Methods for so doing are fairly well known[27] but interests tend to be biased against improved information in this area. As Pack (1981) has pointed out, there can be large private gains (as well as social) from firms investing in acquiring improved information.

The second major area is research and development (R and D) into appropriate technologies. In general, it can be expected that technical change originating in LDCs will tend to be more appropriate than technical change originating in developed countries, since the local environment influences the area of search as well as the results which will prove economically efficient, as illustrated below.[28]

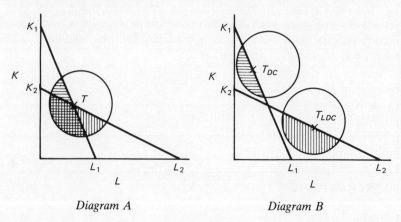

Diagram A   Diagram B

FIGURE 2.1

Suppose R and D tends to be rather randomly dispersed in all directions around the technology in use (roughly a Nelson and Winter type strategy), then for any initial technology, T, in diagram A, new research results will occur within the circle shown. But the only ones that are actually taken up will be those which reduce costs compared with the existing technology – i.e., those below the budget line given by labour and capital costs. It can at once be seen that with DC costs (K_1L_1) the area shaded with horizontal lines will represent economically efficient new techniques. With LDC related relative costs (K_2L_2), the vertically shaded area will represent the techniques that will be developed. Hence techniques developed in DCs will tend to be somewhat more capital intensive than LDC-developed techniques. The difference is, of course, much greater if each start with a different technology, as in diagram B. Here there is no overlapping in the new results. This suggests that the direction of technical change will tend to become cumulative, according to location of research, since the initial differences in new technologies – even after starting with the same technology as in diagram A – will tend to lead to a different choice of technology and consequently to a different starting point for technological developments in the next period.

In the diagram, only the usual capital/labour characteristics have been illustrated. But the actual environment is much broader than this, including the availability of different types of materials, of labour, of various skills, of type of products required and of organisational unit. A major effort then needs to be made to promote LDC technical change. The whole discussion[29] about what does and does not achieve this then becomes relevant. Policies include those directed towards the import of technology, towards the local R and D institutions, towards the promotion of relevant skills, and towards local capital goods industries.

Sources of technology – both as to information and the development of new technology – have been heavily biased towards the modern sector, and within the modern sector towards the relatively large scale. These biases arise from the location of R and D – mainly in developed countries and within developed countries predominantly in the larger scale firms. The informal and traditional sector in LDCs suffers from heavy technical neglect. The same diagram used above to illustrate how developed country technology tends to generate inappropriate techniques for LDCs, could also be interpreted to show how technology change initiated in or from large-scale enterprise in LDCs would tend to generate results inappropriate for the small-scale and traditional sector in LDCs. Hence policies to promote AT should encompass policies to improve sources of technology to the informal and traditional sector as well as policies for improvements towards the modern sector in LDCs.

CONCLUSION

This brief discussion has attempted to identify and classify the main areas in which we should look for macro policies for appropriate technology. It has not discussed the policies themselves in any detail, nor considered the realism or effectiveness of particular policies in particular contexts. Fairly detailed research into comparative country experience with particular policies is necessary to support detailed policy conclusions in each of the areas identified. It is likely that this would lead to a *typology* of countries, with different policies being relevant for countries at differing stages of development and with differing resources.

The general discussion has underlined three conclusions: first, the significance of political economy factors, which must be built into the analysis from the start and not added as an aside at the end; second, the significance of cumulative forces – political, economic and technical – which are such that decisions need to take these into account, and that current options may be severely limited by past decisions; third, the significance of what we have called 'the composition of units', or the proportion of investment decisions in the control of different types of unit. Changing this can, it appears, be of substantial significance with respect to every variable (objectives, resources, markets and technology) and may be one of the most effective ways of promoting AT.

Finally, the dynamic dimension of technology choice needs underlining. Countries are changing, creating resources (for investment in physical capital and human skills); the choice of technique must fit in with this dynamism of the economies, and not simply consist in a static resource allocation exercise. Moreover, technology is also changing at a rapid rate especially in some industries, rendering obsolete many existing technologies. The implications of rapid technology change emanating from developed countries also needs to be taken into account in determining policies for appropriate technology.

NOTES

1. This paper is a revised version of a paper prepared for a *Workshop on Macro-policies for Appropriate Technology*, which took place at the Institute of Social Studies, The Hague, in March 1982 financed by Appropriate Technology International. In preparing this version, I have benefited greatly from the dicussions at the Workshop.

2. Leibenstein (1979), while accepting this definition of 'micro-level' has recently identified a need for a 'micro-micro' economics, exploring how decision-making takes place *within* the decision-making unit. This, too, is an important area for those interested in appropriate technology, and one that has already been approached in discussions of 'engineering' and 'economic' man's relative influence within the firm. See Pickett *et al.* (1974) and Wells (1973). This is not further explored here.

3. This can be seen from a glance at the annual report of almost any of the appropriate technology institutions, which generally consist of a list of particular interventions in relation to technology promotion/dissemination.

4. See, for example, Morawetz (1974), Stewart (1977, chapter 3), Cooper (1979).

5. See Morawetz (1974), p. 517. An almost identical definition is adopted by Westphal (1982).

6. See, for example, Schumacher (1973), Stewart (1977), and the many references in Singh (1981).

7. In Robinson (1979); see Cooper (1979); also Singh (1981) who provides a much fuller discussion of the various characteristics that have been associated with AT.

8. This is not the place to summarise the substantial discussion on the issues of social cost–benefit analysis. See Stewart (1975) and the special issue of *World Development*, edited by Amin and MacArthur (1978).

9. Strictly, the introduction of 'social efficiency' considerations would rule this out, where 'social' values were elucidated with respect to actual government objectives. However, to avoid ambiguity, it is preferable not to eliminate conflicts in this way, so it is assumed that the social efficiency is measured with reference to AT objectives. The use of the term 'government objectives' is ambiguous. In fact there is no unique phenomenon 'government objectives' but rather various objectives belonging to various elements in the governments. Moreover, there is no good way of ascertaining precisely (and sometimes at all) what these objectives are.

10. See Perkins (1983).

11. Galtung (1980), p. 133.

12. I have taken a rather similar view, although expressed somewhat differently. See Stewart (1977, chapter 12).

13. See, for example, Ranis (1973).

14. Enos applies game-theory to choice of technique at a micro level.

15. Formally impossible; see Arrow (1963).

16. See Binswanger's (1978) survey.

17. This happened with respect to refined maize milling in Kenya, for which there was no market until one was 'created', but where subsequently strong demand developed. See Stewart (1977, chapter 9).
18. See Sobhan and Ahmed (1980), Kumar (1982), Perkins (1983).
19. See Stewart (1983).
20. See Acharya (1974), Eckaus (1977).
21. This applies to many craft industries.
22. The arguments for a local capital goods industry for appropriate technology are contained in Stewart (1977, chapter 6).
23. See evidence contained in Forsyth, McBain and Solomon (1980).
24. Sigurdson, at the Hague Workshop.
25. See James and Stewart (1981).
26. There is some evidence that trade in technology between developing countries is more appropriate than North–South trade (see Lecraw 1977); but the (limited) evidence seems to be ambiguous on trade in general. (See Amsden, 1980).
27. See Bhalla (1979).
28. The diagrams are adopted from Binswanger's illustration of the Nelson and Winter model. See Binswanger, Ruttan *et al.* (1978, p. 30).
29. See, for example, Lall (1981), Katz (1978).

REFERENCES

Acharya, S.N. (1974) 'Fiscal/Financial Intervention, Factor Prices and Factor Proportions: A Review of Issues', World Bank Staff Working Paper, No. 183

Amsden, A.H. (1980) 'The Industry Characteristics of Intra-Third World Trade in Manufactures', *Economic Development and Cultural Change*, October.

Arrow, K.J. (1963) *Social Choice and Individual Values*, 2nd ed. (New York: John Wiley).

Bhalla, A. (ed.) (1979) *Towards Global Action for Appropriate Technology* (Oxford, Pergamon Press).

Binswanger, H.P. (1978) *The Economics of Tractors: An Analytic Review* (Hyderabad: ICRISAT).

Binswanger, H.P., Ruttan, V. *et al.* (1978) *Induced Innovation, Technology, Institutions and Development* (Baltimore: Johns Hopkins Press).

Cooper, C. (1979) 'A Summing Up of the Conference', in A. Robinson (ed.) *Appropriate Technologies for Third World Development* (London: Macmillan).

Eckaus, R.S. (1977) *Appropriate Technologies for Developing Countries* (Washington, DC: National Academy of Sciences).

Farvar, T. (1976) 'The Interaction of Ecological and Social Systems: Local Outer Limits in Development', in W.H. Matthews (ed.), *Outer Limits and Human Needs* (Stockholm: Almquist and Wicksell).

Forsyth, D., McBain, N. and Solomon, R. (1980) 'Technical Rigidity and Appropriate Technology in Less Developed Countries', *World Development*, May–June.

Galtung, J. (1980) *The North/South Debate: Technology, Basic Human Needs and the New International Economic Order*, World Order Models Project, Working Paper No. 12.

James, J. and Stewart, F. (1981) 'New Products: A Discussion of the Welfare Effects of the Introduction of New Products in Developing Countries', *Oxford Economic Papers*, March.

Katz, J. (1978) 'Technological Change, Economic Development and Intra and Extra Regional Relations in Latin America', IDB/ECLA/UNDP/IDRC Regional Program of Studies on Scientific and Technical Development in Latin America, Working Paper No. 30 (Buenos Aires: ECLA).

Kumar, R. (1982) 'The Indian Coal Industry After Nationalisation', D. Phil. thesis, Oxford University.

Lall, S. (1981) *Developing Countries as Exporters of Technology* (London: Macmillan).

Lecraw, D. (1977) 'Direct Investment by Firms from Less Developed Countries', *Oxford Economic Papers*, November.

Leibenstein, H. (1979) ' "The Missing Link" – Micro-micro Theory?', *Journal of Economic Literature*, June.

Morawetz, D. (1974) 'Employment Implications of Industrialisation in Developing Countries: A Survey', *Economic Journal*, September.

Morley, S.A. and Smith, G.W. (1974) 'Managerial Discretion and the Choice of Technology by Multinational Firms in Brazil', Paper No. 56 (Rice University).

Nelson, R. and Winter, S. (1974) 'Neo-classical versus Evolutionary Theories of Economic Growth: Critiques and Prospects', *Economic Journal*, December.

Pack, H. (1981) 'Appropriate Industrial Technology: Benefits and Obstacles', *The Annals*, November.

Perkins, F.C. (1983) 'Technology Choice, Industrialisation and Development Experiences in Tanzania', *Journal of Development Studies*, January.

Pickett, J. *et. al.* (1974) 'The Choice of Technology, Economic Efficiency and Employment in Developing Countries', *World Development*, March.

Ranis, G. (1973) 'Industrial Sector Labour Absorption', *Economic Development and Cultural Change*, April.

Ranis, G. (1983) 'Alternative Patterns of Distribution and Growth in the Mixed Economy', In F. Stewart (ed.): *Work, Income and Inequality: Payments Systems in the Third World* (London: Macmillan).

Reddy, A.K. (1979) 'Problems in the Generation of Appropriate Technologies', in A. Robinson (ed.) *Appropriate Technologies for Third World Development* (London: Macmillan).

Robinson, A. (ed.) (1979) *Appropriate Technologies for Third World Development* (London: Macmillan).

Rudra, A. (1982) 'Government Policies for Promoting Appropriate Technology in India', note prepared for Workshop on *Macro Policies for Appropriate Technology*, The Hague, March.

Schumacher, E.F. (1973) *Small is Beautiful: A Study of Economics As If People Mattered* (London: Blond and Briggs).

Singh, H. (1981) 'Appropriate Technology', M. Phil thesis, Oxford University.

Sobhan, R. and Ahmed, M. (1980) *Public Enterprise in an Intermediate*

Regime: A Study in the Political Economy of Bangladesh, Bangladesh Institute of Development Studies.

Stewart, F. (1975) 'A Note on Social Cost Benefit Analysis and Class Conflict in LDCs', *World Development*, January.

Stewart, F. (1977) *Technology and Underdevelopment* (London: Macmillan).

Stewart, F. (ed.) (1983) *Work, Income and Inequality: Payments Systems in the Third World* (London: Macmillan).

Wells, L.T. (1973) 'Economic Man and Engineering Man: A Choice of Technology in a Low Wage Country', *Public Policy*, Summer.

Westphal, L.E. (1982) 'Macro Policies for Appropriate Technology', note for Workshop on *Macro Policies for Appropriate Technology*, The Hague, March.

3 A Game-Theoretic Approach to Choice of Technology in Developing Countries

JOHN L. ENOS

INTRODUCTION: AN APPROACH EMPHASISING POLITICAL ECONOMY

Were there already some generally accredited theory of technological choice in developing countries it would not be necessary to write conceptual papers such as this one: but there exists no theory – that is, no theory whose assumptions exactly conform to conditions in developing countries and which yields hypotheses testable against the data that are available in these countries. What do exist are a group of theories of production related to independent, monolithic private companies – the so-called neoclassical microeconomic theory of the firm[1] – and another theory of the extension of capitalist enterprises from their bases in the developed countries into the developing countries.[2] In this paper we shall suggest that a third body of theory, so-called game-theory, may be at least as useful as these two in describing technological choice in developing countries.

The application of the theory of games to technological choice in developing countries broadens the scope of analysis; it requires that the analyst consider many more economic agents, operating in many more areas, than does neoclassical or Marxist economics. The analyst cannot be content with abstractions such as the 'profit-maximising entrepreneur'; he must consider real people in real situations, with diverse and possibly conflicting interests. He cannot assume that economic agents have perfect knowledge and perfect foresight; he must recognise uncertainty in all its forms. He cannot presume that

47

all decisions are meticulously implemented; he must admit that decisions do not necessarily match outcomes. Finally, he cannot stop theorising when he has laid out the inherited model; he must continually revise his theory as the evidence accumulates.

Such extension of the scope of the analysis, both in complexity and in time, renders this study one of political economy rather than one of narrow economics. It is, of course, economics, in that it focuses on an issue – technological choice – that has conventionally fallen within the purview of economists. But it is also political, in that it considers such non-economic phenomena as influence and power and status, and in that it identifies agents according to their position within various organisations. It is also political in that it recognises that technological choice is not made by a single individual acting autonomously but by a collection of individuals acting in concert.

For the purpose of examining impacts of all these factors and analysing complexities such as conflicting interests held by decision-makers and different roles played by outside organisations (e.g., government), one needs a broad theory like game-theory.

CONCEPTS

There is one concept to be used in this paper which will be familiar to all and several more which will be new to those encountering game-theory for the first time. The familiar concept is that of the 'process'.

The Concept of the 'Process'

The term 'process' is drawn from the language of engineering, where it commonly means a sequence of connected and, ideally, synchronised physical and/or chemical operations. For our purposes, the term will signify a regular sequence of political, economic and technological events whose results are the choice and subsequent development of a new technology. Emphasis will be placed on technological choice and extension within developing countries.

Some of the key attributes of the 'process' of technological choice and development can be stressed. First of all, the process extends through time, taking at the least a few years, and more likely a few decades, to come to completion. It extends across space, too; and across institutions, public and private, profit- and non-profit making, economic and political. Finally it encompasses many individuals, involving some at the very beginning, some later, some at the end, but hardly any single person throughout. It is thus the process that has continuity, and not any of its elements – geographical, institutional, human. It is only the process that can be observed in its entirety, unfolding in time. To focus on a particular period or a particular event is to miss what happened before or afterwards, to focus on one part of the developing country is to miss what happened elsewhere; to focus on one institution or one individual is to miss the contribution of others; even to focus on the new technology itself is to miss the alterations that are obtained through adaptation and improvement. But to focus on the 'process' of 'incorporation' as it might be called, is to include everything within the scope of inquiry.

Another element of the definition of 'process' is its regularity. By being 'regular' the process exhibits the same events in the same historical sequence. From the point of view of the developing country, six different events or stages can generally be observed throughout a complete process of technological incorporation: (i) determining the needs and objectives of the developing country, of the countries that are in a position to supply the new technology, of the organisations that participate in the process, and of the individuals within those organisations; (ii) surveying the alternative technologies and alternative suppliers; (iii) choosing a particular combination of technology, supplier, location, method of finance, etc.; (iv) absorbing the technology in its first application in the developing country; (v) disseminating the technology throughout the economy; and (vi) improving upon the imported technique. A seventh stage may sometimes be observed, namely, conducting research and development within the importing country, leading to a novel and superior technique; but to restrict the scope of this paper we shall limit ourselves to consideration of the first six stages.

There are some implications of focusing on the process of incorporating a technology. Any inquiry into technological choice and development will necessarily extend over several years of a developing country's history. Planning, choice-making, negotiations, construction of plants and equipment, and so on, take time. Moreover, any inquiry will necessitate looking at different institutions and individuals. In the neoclassical microeconomic theory it is one institution, the firm, that undertakes all activities; but in reality it may be one government ministry that plans, and another that chooses the technology and the supplier, and a private firm that constructs the plant and equipment, and yet another private firm that brings it into operation. In the neoclassical microeconomic theory it is one individual, the entrepreneur, that is the sole participant in all the events; but in reality it may be one set of civil servants that plans, and another mixed set of civil servants, politicians and investors that chooses the technology and supplier, and a mixed set of investors, managers and engineers that constructs the plant and equipment, and yet another mixed set that brings it into operation.

The final implication of utilising the term 'process' is that the process may not be brought to completion. Examples of arrested processes are all too common in the developing countries – a plan is made but not implemented, a technology is chosen but not installed, a plant is constructed but not operated: arrests can occur at any of these stages.

Game-Theoretic Concepts

The concepts used in game-theory are now hallowed by nearly forty years of use, so we shall adopt them without change.[3] What is necessary is to interpret them in the light of our subject, so that each customary game-theoretic concept has its analogue in technological choice and development.

The first concept to mention is that of the actor or agent, or 'player' as he or she is called in game-theory. A player is any person, or if monolithic, any group, who influences the outcome. It would be extremely unlikely that the head of the producing firm which employed the technology, the classical 'entrepreneur', would not be a 'player'; but many other individuals, such as ministry secretaries and department chiefs and financiers and possibly even the head of state, could be identified as players too.

So the units of inquiry of game-theory, the players, are not abstractions but elements modelled on the real participants, and their identification is not axiomatic but subjective, depending upon the analyst's view of who has influence upon the outcome. The extent to which any player can influence the outcome will depend, in the real world, upon his power and upon the power of the others in the game. This notion of relative power is formalised in game-theory by means of two additional concepts, 'strategies' and 'pay-offs'. At this point an extremely and unrealistically simple example might help to clarify these two concepts. Imagine a situation in which there are available to the developing country two technologies, which we will label labour intensive and capital intensive, one of which must be chosen. Imagine also that another decision must be made on the rate at which the equipment will be utilised, the alternatives being a low rate of production and a high rate of production. Imagine finally that a binding agreement on the two pairs of choices (labour- versus capital-intensive technique and high- versus low-production rate) must be reached by two players simultaneously. We shall label the two players the Minister of Industry and the Firm Owner, and assume that the former has control over the choice of technology and the latter over the choice of production rate.

Each player has two 'strategies'. The two strategies open to the Minister of Industry are to adopt the labour-intensive technology and to adopt the capital-intensive technology. These are mutually exclusive; only one can be chosen. The two strategies simultaneously open to the Firm Owner are to produce at a low rate of output and to produce at a high rate; he too can choose only one. There are thus four possible outcomes: (i) a labour-intensive technology operated at a low rate; (ii) a labour-intensive technology operated at a high rate; (iii) a capital-intensive technology operated at a low rate; (iv) a capital-intensive technology operated at a high rate. These outcomes are illustrated in Figure 3.1, each of the four alternatives designated by the appropriate cell in the matrix.

The numbers in the cells indicate the returns, or 'pay-offs', assured to each player upon the employment of the strategy. The number in the north-east half of the cell in Figure 3.1 represents the worth to the Firm Owner of following the strategy at the head of the column; the number in the south-west half of the cell the worth to the Minister of Industry. For example, if the strategies simultaneously chosen are the labour-intensive technology operated at a low rate the return to the Firm Owner is +3 and to the Minister of Industry is +3.

Since the matrix of pay-offs in Figure 3.1 will be referred to in the next section on outcomes of games, it might be useful to explain what the numbers representing the pay-offs mean. This explanation will involve discussing both the estimation of the numbers that appear in each cell and also the conditions that are imposed upon the preferences of the players. So far as the estimation of the numbers in the cells is concerned, whereas their absolute values have no significance their relative values were chosen so as to bear some resemblance to the reality of developing countries. The differences in the pay-offs to the Firm Owner, according to which strategy he employs, attempt to reflect typical demand and supply characteristics in internal markets

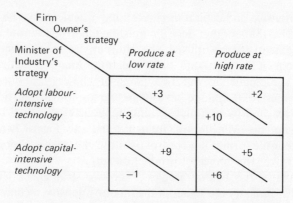

FIGURE 3.1 *Pay-off matrix for game involving Minister of Industry and Owner of Firm in joint decisions on choice of technique and rate of output*

for sophisticated goods, namely, a relatively price-inelastic demand schedule and cost schedules exhibiting mild economies of scale for the labour-intensive plant and substantial economies of scale for the capital-intensive. In addition the capital-intensive strategy is assumed to yield, to the Firm Owner, higher returns than the labour-intensive at both rates of output reflecting the benefits to be gained from a larger allocation of scarce foreign exchange.

The pay-offs to the Minister of Industry are chosen so as to reflect differences in total value-added within the developing country, less that received by the Firm Owner. The higher is the rate of output, the higher would be value-added, so the numbers in the south-west halves of the second column exceed those in the south-west halves of the first column. Moreover, the labour-intensive technology, which is likely to employ more local resources, is assumed to yield greater value-added than the capital-intensive; hence the numbers in the south-west halves of the first row exceed those in the second row. Finally, to devote scarce foreign exchange to purchasing capital-intensive equipment which is then not utilised to its full capacity is assumed to yield a negative return, as the funds might be better used elsewhere in the economy. Consideration of these three factors – rate of output, factor proportions, and opportunity costs – leads to the particular numbers allocated to the Minister of Industry in each cell.

Having described the estimation of the pay-offs it remains to specify the conditions that must be imposed upon preferences, in this example the preferences of the Firm Owner and the Minister of Industry, if solutions to game-theoretic problems are to be obtained.

The pay-offs are the utilities which each individual expects to derive from having played the game; but we shall find it illuminating to consider the pay-offs as money amounts, say so many millions of pesos or so many crores of rupees, and to assume that the players have utility functions that are linear in money. This is equivalent to assuming that an increment in money, say a million pesos, yields the same increase in satisfaction, or is 'worth' the same amount, to each player. Any single amount, appearing as a pay-off, is thereby equally valuable to everyone, and can be considered as transferable utility. Thus, in the north-west cell of the pay-off matrix of Figure 3.1 the +3s which each of the players would receive if they jointly chose to adopt a labour-intensive technology and to operate it at a low rate would be equally valuable to them; whereas the different pay-offs in the north-east cell, +2 to the Firm Owner and +10 to the Minister, would yield the latter five times ($\frac{+10}{+2} = 5$) as much additional utility as the former.

The above game-theory concepts – players, strategies, and outcomes expressed as pay-offs in units of transferable utility – are nearly sufficient for our purposes; the remaining few concepts, having to do with the way in which the players choose, or are forced, to act *vis-à-vis* each other, will be developed in the following section of the paper on methodology.

METHODOLOGY

Two-Person Game-Theory

The objectives of this section on methodology are to select from the entire corpus of game-theory those components that will be most appropriate and to indicate their nearest equivalents in procedures and events arising in the developing countries.

The first component of game-theory that will be described is that explaining the choices of the fewest possible number of interacting participants, so-called two-person theory.[4] Because of its ease in exposition we shall draw upon two-person theory in order to derive three additional concepts – the extent of the rivalry between players, repetition of games, and types of solution. A familiar game, the so-called 'prisoners' dilemma', will serve to illustrate the concepts.

First let us describe the 'prisoners' dilemma'. The game is nearly identical to that of Figure 3.1 – two interacting players, each with an independent and simultaneous choice among two alternative strategies, with the pay-offs, the amounts of money accruing to each of the players as a consequence of their choices, indicated in the four cells. In discussing the game of Figure 3.1 so far we have not asked what the outcome is likely to be; in investigating the alternatives we will now express the dilemma.

Consider first the rational choice of the Minister of Industry, rational that is, in his knowledge of the pay-offs attached to each strategic pair and in his inability to control the other player's choice. In these circumstances he will compare the outcomes for each of *his* two strategies ((a) adopt a labour-intensive technology, or (b) adopt a capital-intensive technology) given that the Firm Owner chooses either of his two ((a) produce at a low rate, or (b) produce at a high rate). The relevant pay-offs are the numbers in the south-west half of each cell in the matrix. For example, if the Minister chooses to adopt the labour-intensive technology he will face two possible outcomes: either the Owner will produce at a low rate or at a high rate. If the Owner produces at a low rate, the outcome for the Minister is +3; if the Owner produces at a high rate, the outcome for the Minister is +10. Obviously, should the Minister choose his labour-intensive strategy, he will hope that the Owner chooses *his* (the Owner's) strategy of producing at a high rate; but he has no assurance that the Owner will so choose. To insure himself against the other player choosing the strategy that produces a worse outcome, the Minister is likely to choose that strategy which will achieve, for him, the best possible outcome *regardless of the other player's choice*.

Considering *both* alternatives on the part of the Owner, the Minister sees that regardless of which one the Owner does choose, he, the Minister, can expect a better outcome if he chooses the labour-intensive technology. The strategy of selecting the labour-intensive technology thus dominates that of the capital-intensive technology; in other words it yields a better result for the Minister whatever strategy is chosen by the Owner.

Carrying out exactly the same comparisons for the Owner, under alternative assumptions as to the choice of strategy by the Minister, reveals that the Owner too has a 'dominant' strategy, namely, to produce at a low rate, being thereby better off whatever the choice of the Minister. And together the two choices determine a unique outcome, the simultaneous decisions of two independent self-interested players, of a labour-intensive technology operated at a low rate. The outcome is indicated by the north-west cell in the pay-off matrix of Figure 3.1, +3 to the Minister and +3 to the Owner.

What is it about this outcome, the result of the choice of two players, each of his dominant strategy, that illustrates the prisoners' dilemma? It is that the outcome in the north-west cell, characterised by the pay-offs (+3 +3), is worse for *both* players than the potential outcome in the south-east cell (+6 +5). Were they to choose the combination represented by the strategies of capital-intensive technology and high output production rate the Minister would be better off, *vis-à-vis* the (+3 +3) outcome by +3 ((+6) − (+3)), and the Owner by +2 ((+5) − (+3)). What is it that prevents them from securing the mutually more advantageous outcome?

The answer does not lie in the assumption of simultaneity of choice. Given the structure of the game it is reasonable that each player should make his choice on the presumption that the other player may choose any of *his* alternatives. To be sure, if one player knows what the other player will choose, in advance of having to declare his own choice, he will be able to select a strategy that will yield an outcome at least as good as that chosen independently. But in the case of the game depicted in Figure 3.1, foreknowledge of the other player's choice does not change the outcome; choices made under the assumptions of simultaneity are just those which would be made sequentially. Such is the conclusion of a competitive game with dominant strategies.

Where the answer does lie is in the assumption of independence, i.e., in the assumption of an absence of co-operation between the two players. All the players would have to do to secure the mutually more satisfactory outcome would be to agree on the strategies that yield the (+6 +5) pay-off pair, and to honour the agreement. Both would be better off. So co-operation enables the players to improve upon the less satisfactory pay-off pair (+3 +3), to escape from the dilemma of the two prisoners in the imaginary game, who are denied mutual communication and consequently cannot maintain their unity in the face of interrogation.[5]

The distinction between independent and united choices of strategies is thus an important one, and is reflected in the terms used to categorise games played under these different conditions – competitive games and co-operative games. Not only is it worthwhile recognising the distinction, but it is also useful estimating the potential gains to both players in shifting from the former to the latter format, so as to provide a measure of the incentive to combine. Finally, it is revealing to calculate the other pair of pay-offs, so as to obtain a measure of the further gain to one player of going back on his word and choosing the non-co-operative strategy, while the other player remains steadfast. Thus, in terms of the pay-off matrix of the game represented in Figure 3.1, the numbers in all the cells are interesting, as any could characterise the final outcome.

When two-person games yielding, for at least one strategic pair, positive pay-offs for both participants are played repeatedly, one would expect the co-operative solution to emerge. Recognition of the potential gains would lead rational players to unite, thereby improving upon the competitive solution.

Identical multiple-play games may therefore produce different outcomes from single-play games, through a shift from competitive to co-operative behaviour. But the attainment of co-operation alone does not enable us to predict the outcome of a game such as that depicted in Figure 3.1. Ambiguous the outcome still is, for there are not one but three possible outcomes – those in the south-west, south-east and north-east cells – all better than the competitive one. To be sure, the outcome in the south-east cell is the only one of the three that yields improvements separately, but the other two yield improvements *in total*. When aggregated, the outcome in the south-west cell yields +8 to the two players (−1 +9), and that in the north-east cell +12, the latter even exceeding the +11 in the south-east cell. Under what conditions might either of these two collective outcomes emerge?

By assuming, as we did in the previous section, transferable utility, we have made possible a choice of the outcome in the north-east cell (yielding +12 in total) to that in the south-east cell (yielding +11); but this assumption is not sufficient. To secure the outcome an additional assumption is necessary, that the player who does substantially better by the shift from the co-operative solution, in the south-east cell, to the 'integrated' solution, in the north-east cell, compensate the player who would otherwise do worse. In the example of Figure 3.1, the co-operative solution yields +6 to the Minister and +5 to the Owner, whereas the integrated solution yields +10 to the Minister and only +2 to the Owner. To persuade the Owner to produce at a high rate in a *labour-intensive* plant, the Minister would have to subsidise him by at least +3 (the difference between +5 and +2). Since the Minister gains +4 (the difference between +6 and +10) by the shift, he would have the wherewithal for the subsidy. Our theory cannot tell us exactly how much the subsidy will be, only that it will lie between +3 (the least the Owner would accept to shift) and +4 (the most the Minister would pay). Game-theorists use the term 'side-payments' to describe such subsidies or transfers which are outside the context of the game itself. There is nothing in the structure of the game as depicted in Figure 3.1 that admits a subsidy; if it occurs it comes about through bargaining among the two players, followed by the joint choice of the strategic pair that yields the largest total pay-off.

This discussion of side-payments concludes our remarks on two-person games. We have introduced several additional concepts, chiefly those of competitive, co-operative and integrated games, and of the associated negotiations and strategic choices that generate their logical outcomes. When the two players come to decisions independently, without any foreknowledge of their rival's choice, they may each find that one strategy dominates the alternatives and determines a unique outcome. When, by the rules of the game, the players are allowed to co-operate, a different outcome may emerge, one that is mutually preferable to the competitive solution. Finally, when the players are not only allowed to co-operate but also to make side-payments, that outcome which yields the highest total pay-off, the integrated solution, may result.

These three outcomes – the competitive, the co-operative and the integrated – are all illustrated in the various cells of the pay-off matrix of Figure 3.1: the competitive solution (+3 +3), the co-operative (+6 +5) and the integrated (+10 +2, with a subsequent side-payment of between +3 and +4). But there remains in Figure 3.1 one possible outcome, that resulting from the choice of a capital-intensive technology by the Minister and a low rate of output by the Firm Owner, yielding a pay-off of (−1 +9), that has not yet been investigated. Although this outcome would not reasonably emerge as a consequence of any of the three modes of decision discussed so far, anyone who has observed the actual choice of techniques in developing countries will recognise that it is an outcome all too often encountered. But such an outcome cannot be deduced from two-person game-theory; it is necessary to expand the number of participants to allow the entry of additional players. The extension of game-theory to more than two, or as it is commonly called n-persons, occupies us next.

Extensions of Two-person Game-Theory to Account for Many Players

Anyone who has read this far may well have wondered if the game depicted in Figure 3.1 could truly be said to involve only two players, for there is implicit some third person who is providing the positive pay-offs. This absent benefactor, who is commonly called 'Nature' by game-theorists, should be considered explicitly whenever the algebraic sum of the pay-offs in one or more cells of the matrix differs from zero; i.e., whenever the game is non-zero sum.

Nature's ambiguous role in two-person, non-zero sum games arises from the fact that it, unlike the identified players, has no strategic choice. In the game of Figure 3.1, Nature has no means of assuring that there materialises the competitive outcome involving the least total pay-off (+3 +3); it has no means of preventing the integrated outcome involving the greatest total pay-off (+10 +2). Yet to give Nature an active rather than a passive role will obviously complicate the theory: is it worth doing?

To anyone attempting to apply game-theory to technical choice in developing countries the answer must be affirmative, because what the game-theorist innocuously calls Nature is, generally, in the context of the developing countries, the public interest. Moreover, the public interest is not necessarily synonymous with, nor is it necessarily opposed to, the interests of the individuals participating in the choice of technology. In the game underlying Figure 3.1 we assumed that the interest of the Minister of Industry and the public interest were identical, but this was not a statement of fact, nor a general tendency, but a simplification, in order to reduce the number of players to two. If, as happens often in the real world, the interest of the Minister of Industry and his officials diverges from the public interest, the representation as a game involves three players and the combined pay-off of the original two players – the Minister and the Firm Owner – need not bear any resemblance to the pay-off of the third. A neoclassical economist, believing in the 'hidden hand' that equates the private and the public interests, might assume that the pay-off of the Firm Owner was congruent with the pay-off of the public; but the development economist, having observed all too many cases where there is no congruence, would not be surprised to discover negative pay-off for the public in the cells in which there was positive pay-off for the Firm Owner or vice versa.

Allowing explicitly for the public interest, therefore, extends the number of players beyond two, complicating the theory greatly. It is not simply that the dimensionality of the game increases from two to three, nor that the pay-off matrix becomes a cube rather than a rectangle, but that allowances must be made for the formation of coalitions and for their influence on possible outcomes. In the previous section we did consider coalitions implicity, when we allowed the Minister and the Firm Owner jointly to choose the more favourable pair of strategies resulting in a capital-intensive technology operated at a high rate. Two-person co-operative games can be thought of as coalitions against Nature. But now we afford the coalition's opponent an active role too, meaning a choice among alternative strategies and an opportunity to ally himself with one or more of the other players.

The additional complexity of the game can be illustrated by considering the number of different possible coalitions with three players, now labelled for simplicity M, F and P. First of all, the three players can each act independently; this eventually can be described as three one-person coalitions. Secondly, M and F can ally themselves against P, forming a two-person coalition. But equally possibly M and P can form a coalition against F; or F and P against M. Finally, all three players can form a triple alliance. Thus there are five possible sets of allegiance (M, F and P choosing strategies independently; M and F co-operating; M and P co-operating; F and P co-operating; and M, F and P co-operating) to be allowed for, with their different implications for strategic choice. In game-theory as in the real world the observer confronts alignments made to advance mutual interest.

In the real world single solutions emerge; in three-person game-theory, where pay-offs are ascribed to every possible trial of strategies, some device is necessary to generate a single outcome, some solution concept. Several have been suggested; the two that we will focus on are the concepts of the 'core' and the 'bargaining set'. The core is the more restrictive of the two, being defined as that set of imputations which are not dominated by any coalition which could potentially be established. 'Imputation' is a technical term designating the entire distribution of costs or benefits (in the form of transferable utility) among all the players, such that each player acting singly or as a member of some coalition receives at least as much as he could obtain by acting independently, and that all players, together, receive the maximum amount that 'Nature', or the game, provides. To determine whether or not a possible solution – i.e., a set of pay-offs to each player, plus any side-payments – lies in the core one must examine every possible size and combination of coalitions, and their imputations. Only those possible solutions which cannot be blocked, via the formation of a counter-coalition promising higher rewards to its potential members, constitute the core.

There are, from our point of view, two limitations to the solution concept of the core. First of all, the amount of calculation, communication and negotiation that would be required of the participants in any game in the real world would be far beyond their abilities. We saw in the case of three players how many possible coalitions would have to be contemplated and evaluated; in the case of four, the number would increase from five to fifteen. For players more numerous than four, the number of coalitions, and combinations of coalitions, to be tested would be beyond the realms of possibility.

The second limitation to the solution concept of the core lies not in the impossibly wide range of alternative coalitions to consider but in the overly narrow range of solutions that may actually comprise the core. Most, if not all, of the feasible solutions may be eliminated by counter-coalitions. For example, if the total in each cell of the pay-off matrix is the same (so-called 'constant-sum' games), the core will be empty for all games with three or more players. An empty core indicates that there is no outcome that all the players will willingly submit to.

To surmount these limitations one may appeal to a second solution concept, that of the 'bargaining set'. The bargaining set is a more elastic concept than the core, for it contains those outcomes which would otherwise be contained within the core were they not capable of being blocked by some other, potential coalition. The bargaining set thus includes imputations associated with all feasible coalitions which would yield to each member a pay-off greater than he or she would obtain by acting alone. All coalitional structures that would provide positive benefits to their members, regardless of the opposition, constitute the bargaining set.

It is probably obvious that the core, and even more so the bargaining set, may contain multiple solutions. To the economic theorist, interested chiefly in the existence of unique outcomes, such multiplicity is displeasing, because it makes it impossible to deduce limiting principles. To the applied economist, however, the uniqueness of solutions is not necessarily to be sought for, and is quite disadvantageous if the theoretical process of reducing alternatives to a single choice eliminates outcomes that might arise in the real world. We saw in the game underlying Figure 3.1, for example, that at least three of the four possible outcomes indicated in the pay-off matrix could occur; a solution concept (e.g., that leading to the outcome labelled the prisoner's dilemma) which eliminated two of those three, so as to derive a unique outcome, would miss much that is relevant to a study of the choice of technology in developing countries. Ambiguity is desirable in theory if it reflects accurately ambiguity in reality.

In the following section we shall be concerned to apply the above terms and solution concepts to the analysis of technology choice. At this stage, however, it is useful to note that the choice of technology can easily be seen as analogous to part of the outcome of the game. The choice of technology is a discrete, identifiable decision, with major effects upon such subsequent phenomena as expenditure of foreign exchange, employment and the distribution of income and wealth. Choice of technology is not the entire outcome, of course, for still to be determined are such matters as the speed and efficiency of construction, the degree of local absorption of the new techniques, the rate at which the equipment is operated, and so on. These matters do not fit so well into the game-theoretic framework, not usually being decided simultaneously with the choice of technology, nor being discrete and easily observable. In consolidating these alternatives with the technological alternatives, so as to obtain a single pay-off matrix, we are uniting the problematic with the certain, the extended with the once and for all. Yet the format of game-theory, as we have developed it, requires a unique and consolidated pay-off matrix, if outcomes are to be explained. The morals to be drawn are that the consequences of the choices made, represented in the pay-off matrix by the numbers in the cells, cannot be predicted with any great precision; and that the actual choices made by those participants who decide later events can diverge from the choices they promised earlier. In the language of game-theory, this latter qualification concerns the 'enforceability' of a simultaneous choice; in our example it appeared when we admitted that the Firm Owner could renege on a promise to operate the chosen equipment at a high rate. In the next two sections of analysis and experience we will have to keep both these morals in mind, enquiring repeatedly into the accuracy of the pay-off figures and the enforceability of the choices.

ANALYSIS

Characterisation of Choices Frequently Observed in the Developing Countries

In this section we shall attempt to characterise the choices of technology that are made in developing countries, devising categories into one or another of which most actual choices can be fitted and describing these categories in game-theoretic terms. The purposes of the analysis are three-fold, to create a classification scheme that will be clear and more or less all-inclusive; to identify what alternatives there are to the choices actually made; and to formulate an ideal, as a standard against which actual choices can be judged.

Meeting the first objective of classification will involve summarising choices of technology in such a way as to yield a few mutually exclusive groups or categories into one of which any historical choice can be placed. Two such summary indicators have already been used in the construction of the game illustrated in Table 3.1; namely, capital- versus labour-intensive processes and high versus low rates of output. Five others have been mentioned in the exposition of the previous sections; namely, the numbers of participants, their degree of co-operation, the extent of their knowledge of each other's strategies, the simultaneous or sequential nature of decisions, and the adherence or lack of adherence to promises. With these seven indicators we will assemble four major categories, which together will incorporate most actual choices.

The first major category will be called the overly capital-intensive choice, an outcome that is discussed endlessly in the literature on technological choice and that has been frequently encountered. Its main indicator is a capital–labour ratio far higher than the availability of inputs, both contemporary and future, would recommend; other indicators are relatively few participants co-operating in the choice of technology. The second category will be called the centralised or bureaucratic choice, in which the technological decisions are made by a single body, either a government ministry or a corporation, public or private. Additional characteristics of the bureaucratic choice are a quite limited knowledge of the environment within which the process incorporating the technology will operate and, partly as a consequence, an inability to operate the equipment at its full potential.

The third category is the fragmented choice, in which sequential decisions are spread among many actors. Knowledge of the strategies of other participants is often limited, and co-operation, in the forms of both combined negotiations and fulfilment of commitments, is frequently lacking. Again, as a consequence, the full potential of capital equipment and other inputs may not be realised.

The fourth and final category is that of conspiratorial choice, implying that two or three individuals form a coalition which operates clandestinely in order to secure private gains. The interests represented are solely those of the participants, and the general interest may well be compromised in the choice. The decisive individuals co-operate closely, assuring for themselves the outcome with the

TABLE 3.1 Indicators of 'stylised' outcomes in the choice of technology

Indicator	Outcome	Overly capital-intensive choice	Bureaucratic choice	Fragmented choice	Conspiratorial choice
Capital/labour ratio	Capital intensive	X			X
	Labour intensive				
Rate of output	Low		X or	X	X or
	High		X		X
Number of active participants	Two	X or	X		X or
	Three	X			X
	More			X	
Degree of co-operation	Little				
	Co-operative	X			
	Integrated				X
Extent of knowledge	Limited		X	X	X
	Nearly complete				
Temporal speed of decisions	Simultaneous			X	X
	Extended				
Adherence to promises	Little			X	X or
	Some				
	Complete commitment				X

highest total pay-off by making any necessary side-payments. The technological choice is almost always that of a capital-intensive process, which may subsequently be operated at close to full capacity if the coalition maintains its unity. If the coalition dissolves, however, the consequence is likely to be recrimination and stagnation of output.

All four 'stylised' outcomes – the overly capital-intensive choice, the bureaucratic choice, the fragmented choice and the conspiratorial choice – are identified in Table 3.1 by their chief characteristics. Where there is an 'X' in the column indicated by the category it means that the indicator in the appropriate row is relevant to the description. Blanks in the column mean either that the indicator is not significant or that no generalisation can be made. There may well be other stylised choices, in addition to those identified; it is easy to define additional categories simply by putting a series of 'X's in columns that are empty in Table 3.1. For example, one could create a fifth category, calling it, say, propagandist choice, by putting 'X's in under the indicators capital intensive, two participants (usually the foreign donor and the national leader), integrated and simultaneous decisions and low rate of output. One could create many more too; but our purpose is not to exhaust the number of possible categories of technological choice but rather to demonstrate the usefulness of game-theoretic concepts and models. In limiting our categories to four we have, we hope, kept the number of analytical alternatives within bounds and yet succeeded in characterising reasonably accurately the majority of technological choices observed in developing countries.

It will be informative to try to complete the game-theoretic descriptions of the situations which lead to one of the four stylised outcomes of Table 3.1. This task will require adding more dimensions to the schema in Table 3.1, and will reveal something of the alternatives to the stylised outcomes.

An Analysis of the Overly Capital-Intensive Choice

Let us take the first stylised outcome and begin by filling in the outlines of the overly capital-intensive choice. The number of players is not likely to exceed three, consisting generally of the administrator of a foreign assistance programme of a developed country, an official from the appropriate ministry in the developing country and, if the project is to fall within the private sector, the owner of the firm that will undertake production. The three players will have quite good ideas of the strategies available to themselves and to the other players, and of the immediate pay-offs associated with each possible outcome. The rules of the game encourage co-operation between the players in arriving at a joint decision, and usually permit the formation of coalitions, occasionally of all three players, more often of any two players.

Since the pay-off matrix of a three-person game has three dimensions, one for each of the three players, its display will be more complex than the two dimensional matrix of Figure 3.1. Even limiting the strategic alternatives to two to each player, as is done in Figure 3.2, still yields eight possible outcomes. The pay-offs associated with these eight possible outcomes are provided in Figure 3.3, the left-hand sub-matrix being appropriate for the choice of the overly capital-intensive technology and the right-hand sub-matrix for the choice of the less capital-intensive technology. Each sub-matrix in Figure 3.3 provides four sets of pay-offs, each set attributable to different strategic choices on the part of the three players; together the two sub-matrices display all eight possible outcomes. The 'stylised' outcome – resulting from a combination of simultaneous choices of conducting speedy negotiations by the Ministry, granting generous terms by the Aid Agency, and selecting the overly capital-intensive technology by the Firm Owner – is located in the upper left-hand cell in the left-hand sub-matrix in Figure 3.3; the pay-offs accruing to the players are +3 to the Ministry, +3 to the Aid Agency, and +5 to the Firm Owner.

These and the other seven sets of pay-offs do not come out of thin air: they must be estimated on the basis of the situation within which the game is played. This exercise of estimation involves identifying the strategies available to each player, and evaluating all of the alternative outcomes from the point of view of each player. To the Aid Agency of the developed country, the strategies are assumed to be to grant assistance on either lenient or severe terms, the latter being the preferred alternative partly because it enables a fixed allocation of funds to be extended over more projects. To the Ministry in the developing country the strategies are to conduct negotiations either quickly or slowly, the former being preferred because it economises on scarce administrative resources. To the Firm Owner the strategies are to pick an overly capital-intensive technology or a less capital-intensive one, the latter being preferred, we assume, because production costs would be lower.

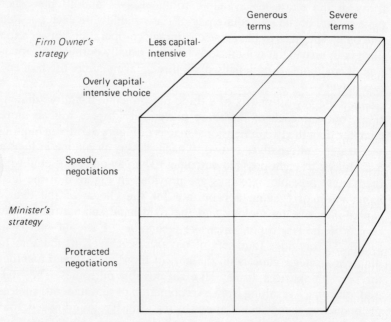
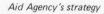

FIGURE 3.2 *Pay-off matrix for the overly capital-intensive choice*

Given
Firm Owner chooses overly capital-
intensive technology

Given
Firm Owner chooses less capital-
intensive technology

Aid Agency's strategy ⟍ Minister's strategy	Generous terms	Severe terms	Aid Agency's strategy ⟍ Minister's strategy	Generous terms	Severe terms
Speedy negotiations	+3 ⟍ +3 (+5)	+5 ⟍ +1 (+2)	Speedy negotiations	+1 ⟍ +1 (+6)	+3 ⟍ 0 (+3)
Protracted negotiations	+1 ⟍ +2 (+1)	+2 ⟍ 0 (0)	Protracted negotiations	0 ⟍ +5 (+4)	+1 ⟍ +3 (+2)

Note: Pay-off matrices for the Ministry and the Aid Agency (the Firm
Owner's pay-offs in parentheses).

FIGURE 3.3 *Partitioned pay-off matrix for the*
overly capital-intensive choice

Evaluating alternative outcomes is quite a complex task, involving as it does the determination of eight sets of three pay-offs each. We will illustrate the estimations for two sets, leaving the determination of the other six implicit. It is not that the estimates are obvious, nor that someone else would obtain the same numbers in the cells, but that we do not have enough space to record all the estimates: the two examples will have to suffice. The first set to be estimated is the 'stylised' outcome, resulting from the simultaneous, co-operative choices of generous terms by the Aid Agency, speedy negotiations by the Ministry, and overly capital-intensive technology by the Firm Owner: and the second set is the antithetic outcome of severe terms, protracted negotiations and a less capital-intensive technology. The first set is depicted in the upper left cell of the left-hand matrix in Figure 3.3, the second set in the lower right cell of the right-hand matrix.

The differences in the pay-offs in the two cells stem chiefly from the presumption that the overly capital-intensive technology is already being employed in the developed country granting the foreign assistance. Its Aid Agency would prefer this technology to be adopted without alteration, for its own national producer could supply the design 'off the shelf', quickly and profitably. To adapt the design to the resource mix of the client firm in the developing country would take time, would involve the Agency in extensive collaboration with the supplying firm, the using firm and the Ministry. The Ministry would prefer the technology to be adapted to local conditions, as that would conserve scarce foreign exchange, and would even be willing to extend considerable time assuring that the adaptation was carried out; nevertheless it can see the merits of a speedy acceptance and installation of the existing highly capital-intensive technology, particularly if the Aid Agency will make this available on more generous terms. The Firm which will utilise the technology in the developing country is less concerned about the capital-intensity of the process than it is about the time taken to come to agreement with the Aid Agency and the Ministry and with the terms on which the plant and equipment are supplied.

Given these considerations, the 'stylised' outcome is estimated to yield pay-offs of +3 to the Aid Agency, +3 to the Ministry and +5 to the Firm in the developing country, compared to pay-offs of respectively +1, +3 and +2 for the outcome associated with the adaptation of the technology to a less capital-intensive form, available on more severe terms and requiring protracted negotiations. To the Ministry, these two outcomes are more or less equally attractive; but to the Aid Agency the latter requires more attention and is likely to offend the supplying firm, and to the adopting Firm the latter requires inordinate and costly delays. The less attractive nature of the latter outcome is reflected in the lower pay-offs (+3 less +1 or 2; and +5 less +2 or 3) assigned to the Aid Agency and the Firm Owner respectively.

Similar arguments are used to estimate the pay-offs in the other six cells of the sub-matrices. Looking now at the 'stylised' outcome, in comparison to *all* its alternatives, we see that the 'stylised' outcome has two attributes: first, its pay-offs add up to a larger total (+3+3+5 = 11) than do those in any other cell (the remainder range from 0+5+4 = 9, resulting from the choice of a less capital-intensive technology acquired on generous terms after prolonged consultation, to 2+0+0 = 2, resulting from the choice of the overly capital-intensive technology acquired on severe terms after protracted negotiations). Secondly, the 'stylised' outcome is not the most attractive alternative, when considered separately, for any of the three participants. For the Aid Agency, the single most attractive outcome, yielding a pay-off of +5, is that associated with the choice of the overly capital-intensive technology, after speedy negotiations, and yet supplied on the more severe terms. For the Ministry, the single most attractive alternative, yielding a pay-off of +5, is the already familiar one of an appropriate technology developed through mutual efforts and made available on generous terms. For the Firm, yielding a pay-off of +6, the single most attractive alternative is also the already familiar one of the appropriate technique provided without delay and on generous terms.

What is it about the 'stylised' outcome that makes it the likely choice of all three actors, each of whom would prefer another? One reason has already been provided, namely, that the 'stylised' outcome generates the highest total rewards to the three participants combined. Acting collectively, they can do no better. The second reason is that for each participant, the 'stylised' outcome ranks higher than all the six remaining alternatives. Once the single most attractive alternative has been abandoned, the 'stylised' outcome is everyone's 'second best'. The 'first best' for each participant taken separately would arouse considerable opposition on the part of one of the other participants, as can be seen by examining the pay-off in the upper right-hand cell of the left-hand sub-matrix (with pay-offs +5, +1, +2), the upper left-hand cell in the right-hand sub-matrix (with pay-offs +1, +1, +6), and the lower left-hand cell of the right-hand sub-matrix (with pay-offs 0, +5, +4). The 'stylised' out-come would not be so strongly opposed by any participant because he would secure nearly as much as the best conceivable and far better than the worst. Each gives a little in the course of arriving at a joint decision, and the overly capital-intensive choice emerges.

What have we obtained from the game-theoretic exercise besides an explanation of the likely choice in the situation described? The first additional piece of information is that there is not one aspect to the 'stylised' choice but three. To be sure, the technology chosen was the overly capital-intensive one, but at the same time the Aid Agency chose to make the technology available to the developing country on generous terms and the Ministry chose to expedite the negotiations. We have learned that choices are multidimensional.

The second additional piece of information is that the co-operative choice led to the maximum benefits for the participants as a group, and to near-maximum for the participants singly. For all it was the most, for each the 'second-best'.[6] Joint negotiations in which each participant sacrifices a little in order to achieve a harmonious out-come yields, in such a situation as depicted in Figure 3.3, a solution superior to any alternative. By contrast, the alternatives which yield very low total pay-off, for example, the one in the lower right-hand cell of the left-hand sub-matrix of Figure 3.3 (+2, 0, 0), are the type which could emerge only after prolonged and acrimonious negotia-tions, by which time the participants were willing to accept worse outcomes in order to inflict losses on the others. These are the analogues, on the scene of international bargaining, of combined strikes and lockouts in industrial relations: the co-operative nature of our game formulation prohibits such mutually destructive results.

The third additional piece or pieces of information we obtain from the game-theoretic analysis is the 'distance' of non-chosen alterna-tives from the 'stylised' choice. Were there any alternatives that were nearly as attractive: and, if so, which were they? How much would each participant sacrifice if a different alternative were chosen in-stead? Or, putting the question another way, by how much would each participant have to be subsidised to shift the co-operative choice to a specified alternative?

With the pay-off sub-matrices of Figure 3.3 we can answer these questions. The alternative closest to the 'stylised' choice, where distance is measured in the simplest way possible, namely, by com-paring total pay-offs, is the choice indicated by a less capital-intensive technology developed with collaboration and supplied on generous terms. This alternative, which we might call 'development of the appropriate technology', yields the maximum (+5) pay-off for the Ministry; a substantial, although not maximum (+4) pay-off to the adopting Firm; and the lowest pay-off (0) to the Aid Agency. The arguments underlying the estimation of the first two pay-offs have already been given, the argument underlying the estimation of the pay-off of zero for the Aid Agency is that the Agency incurs the resentment of the supplier, who is forced to allocate resources to research and development in the course of adapting the technology to the resource mix in the developing country, and yet grants concessions to the Ministry and the importing Firm, in the course of specifying the nature of the adaptation and of financing the plant and equipment.

There are other possible ways of measuring 'distance', however, which yield different 'runners-up' to the 'stylised' choice. One measure which would draw some support on theoretical grounds would be that which minimised the sum of the squares of the differences between the pay-offs to each participant in the 'stylised' choice and in the alternative under consideration, or minimising the so-called 'quadratic loss function'. Under this criterion, the nearest alternative is the choice indicated in the upper left-hand cell of the right-hand sub-matrix (+1, +1, +6), equivalent to a less capital-intensive technology developed after little consultation with the Ministry but supplied to the importing Firm on generous terms. Such a choice might be called 'by-passing the local authorities'; it yields the maximum pay-off to the importing Firm, but lesser pay-offs to the Aid Agency, which now earns the resentment of both the supplier and the Ministry, and to the Ministry, which loses the opportunity to assure that the national interests are taken account of in the adaptation of the technology.

Contrasting these two runners-up, we might guess that the first, or 'development of the appropriate technology', would be more beneficial to the developing country than the second or 'by-passing the local authorities', provided that including the Ministry in the adaptation of the technology did not consume an inordinate amount of time. But as to which alternative would be more feasible, more easily obtained if the 'overly capital-intensive choice' were to be avoided, it is almost impossible to guess. The Aid Agency would much prefer the 'stylised' choice to the first alternative, and the Ministry the 'stylised' choice to the second. It is rather unlikely that either the Aid Agency or the Ministry would accept being excluded from the process of developing the appropriate technology, and it is difficult to imagine a mode of persuasion. Given the prospect of the Ministry choosing to prolong negotiations and the importing Firm choosing to accept a less capital-intensive technology, the Aid Agency might well seize upon the strategy of imposing severe terms, which would shift the outcome to the lower right-hand cell of the right-hand sub-matrix and leave both other participants worse off; given the prospect of the Aid Agency choosing to grant generous terms and the Firm to accept a less capital-intensive technology, the Ministry might well seize upon the strategy of prolonging negotiations, which would shift the outcome to the lower left-hand cell of the right-hand sub-matrix and leave both other participants worse off. Neither of the runners-up seems to be a stable solution, unlike the stylised choice itself. Thus we have picked up a fourth additional piece of information, namely, that in the situation described, the stylised solution is, of the three most attractive alternatives, the more nearly stable outcome. The 'overly capital-intensive choice' alone is sufficiently attractive to all players to recommend itself as their joint decision.

In game-theoretic terms we have characterised the overly capital-intensive choice as a three-person game with full information and with two strategies available to each player. The rules of the game require simultaneous decisions and permit co-operation, but the given institutional setting does not encompass side-payments. The most likely outcome is, for each of the players separately, the second best, but for all together it is the alternative that maximises total returns. Were side payments permitted, we would say that the stylised outcome is in the bargaining set of the two-player coalitions comprised of the Aid Agency and the Firm Owner, but is not in the core, for it can be blocked by a proto-coalition comprised of the Ministry and the Firm Owner. In fact, in this game the core is empty. But the 'rules' did not permit side-payments, meaning that the game whose stylised outcome is the overly capital-intensive choice is one in which each player is bound to respect the interests of his constituency. In the case of the Firm Owner, whose constituency is himself, this is not a serious restriction, but in the cases of the Aid Agency and the Ministry it can be. It does happen that the Aid Agency subordinates the interests of the developed country to those of its own bureaucracy or to those of the companies supplying techniques, and that the Ministry subordinates the interests of the developing country to those of its own officials; in these situations the likely outcome falls not in the category designated the 'overly capital-intensive choice' but by that designated the 'conspiratorial choice'. Game-theory makes no moral judgements, nor even economic judgements; it merely suggests that the rules of the game applying in 'conspiratorial choice' are those permitting the formation of coalitions with side-payments.

The possibility that there could be a divergence of interest between those of the participants in the decisions and those of the developing country as a whole cannot be denied. Game-theory cannot reconcile the conflict of interests but what it can do is suggest a standard against which the 'stylised' outcomes can be measured. It is to the derivation of a standard, or ideal, that the remaining portion of this section will be devoted.

The Characteristics of the Ideal Choice

There is a final category of 'stylised' outcomes that we should like to establish, namely, the 'ideal choice'. Every choice of technology has an impact upon the economy of the developing country and so involves the national interest. It is not controversial to argue that the national interest should be considered explicitly, nor is it controversial to try to create a set of criteria whose attainment would guarantee that the national interest were being respected. What is controversial is to suggest a specific set of criteria, and to elevate these criteria above the alternative sets of other economists. To the extent possible we will restrict ourselves to the less controversial aspect – the attempt to create a set of criteria that would guarantee respect for the national interest – avoiding the more controversial aspects of why these criteria are desirable and how they might be imposed.

Controversy will not be reduced, although it may be camouflaged, by expressing the criteria in game-theoretic form. This means characterising the ideal in terms of the numbers and types of players, the conditions of play and the outcomes.

Since it is the results that we are ultimately concerned with, let us proceed in our citation of the ideal by considering the outcome first. What is ideal from the point of view of the developing country as a whole? No generally acceptable definition is close at hand, but one that might enlist support is that outcome which enables the output associated with the project to be produced at the lowest cost, when all inputs are valued at their shadow prices.[7] Such an outcome would be indicated in Figure 3.3 by the highest pay-off ascribed to the

Ministry (+5 in the lower left-hand cell of the right-hand sub-matrix, consequential upon the choice of an appropriate technology granted on generous terms), for one of the assumptions underlying the game depicted in Figure 3.3 was that the Ministry in its negotiations faithfully represented the national interest. (Such an outcome was not represented by the game in Figure 3.1, whose players were all acting in their own rather than in any social interest.)

Picking a measure of the best outcome is only the first step in the creation of the ideal category; the second step is to associate the most desirable outcome with possible game-theoretic structures and rules, and the third to assure consistency between each given structure and set of rules on the one hand and the most desirable outcome on the other. In mathematical language what we seek is a one-to-one transformation from structure and rules to outcome, not a one-to-many.

Two possible combinations of structure and rules suggest themselves, one derived from economic theory and the other from recent economic history. The theoretical edifice that generates the former combination is neoclassical microeconomics, specifically its perfectly competitive paradigm.[8] In game-theoretic language the perfectly competitive outcome requires an infinitely large number of participants, playing under any rules, non-co-operative, co-operative or integrated. Even though they may, according to the rules, form coalitions, it turns out that if the number of players is large enough no coalition guarantees more to its members than they would each receive acting alone. Using game-theoretic language, we say that as the number of players increases, in an orderly fashion, the core of solutions shrinks, converging in the limit to the perfectly competitive solution.[9] Since the perfectly competitive solution is unique and, under plausible conditions, stable, it provides the one-to-one transformation that is required. The difficulty with the perfectly competitive model as an ideal is not its results nor its logic but its requirement of an infinitely large number of participants. No decision governing the choice of technology in a developing country, no game in reality, has an infinite number of players.

Many economists, if asked to give examples of developing countries, which have fairly consistently made the right decisions on the choice and application of technology, would reply with names such as Singapore and the Republic of Korea. Decision-making in these countries is centralised, the chief actor being the appropriate government ministry.[10] The Ministry's objective is consistent with our ideal of minimum cost production and the Ministry's actions are consistent with their objective. Negotiations on each project are initiated with many foreign suppliers, in the case of the Republic of Korea generally with everyone capable of supplying the appropriate technology. As the terms under which the foreign suppliers may participate become more severe, more advantageous to the developing country, the potential suppliers drop out, until finally only one remains. Agreement is then reached on terms most advantageous to the importing country. Finally, by means of continued scrutiny of the project and punishments for violations of the terms, the Ministry assures that the agreement is implemented.

Using game-theoretic language we would say that the players are numerous and consist of the Ministry and many prospective suppliers of the foreign technology. The rules of the game forbid collusion between any foreign supplier and the Ministry: i.e., the game is non-co-operative. Knowledge of opponents' strategies is nearly complete. The Ministry's choice is that strategy which yields it the highest pay-off, pay-offs being inversely proportional to production costs of the imported process. The successful foreign firm plays the only strategy open to it, namely, to accept the assignment: it is successful in winning the contract because it is the only player which can obtain a positive pay-off, given the strategy adopted by the Ministry: all other players would receive negative pay-offs and hence refuse to play the game. The particular features of the ideal game that secure the most attractive outcome for the developing country are the multitude of players forced to compete with one another for the Ministry's favour, full knowledge in the hands of the Ministry, and complete identification of the Ministry's interests with those of the developing nation at large. Relax any of these conditions and the complexion of the game changes, with the consequence that the ideal outcome is no longer assured.

Devising the pay-off matrix for the above game is relatively easy, although its multidimensional form prevents displaying it in full. What can be displayed is the sub-matrix involving the Ministry and the successful foreign supplier, shown in Figure 3.4. (Sub-matrices involving the Ministry and all other foreign suppliers would have negative pay-offs for the latter in the top left-hand column.) Looking at the pay-off matrix from the point of view of the Ministry we see that choosing the most attractive terms for the developing country is a dominant strategy, yielding more desirable outcomes no matter what the supplying firm's choice. The same is true for the supplying

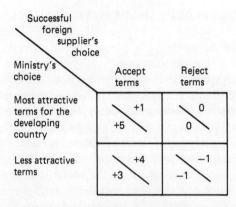

FIGURE 3.4 *Pay-off matrix for the 'ideal choice'*

firm, whose strategy 'Accept' dominates 'Reject'. Hence, in the non-co-operative game, the 'ideal choice' (+5 +1) is assured.

Not only does the pay-off matrix of Figure 3.4 enable us to deduce that the 'ideal choice' can emerge, but it also points out the crucial importance of the rules under which the game is played. What if the game were integrated rather than non-co-operative? A comparison of the two cells in the left-hand column of Figure 3.4 reveals that in the lower cell (+3 +4 = +7) the total pay-off is greater than in the upper (+5 +1 = 6). There would be more to share out between the Ministry and the successful Firm if the former chose to impose terms less attractive for the developing country. Such an outcome could result in practice if the supplying Firm were willing to make a side-payment greater than +2, say +2.5, to the Ministry. As a consequence the Ministry would receive more (+3 +2.5 = +5.5, exceeding the +5 of the 'ideal choice'), as would the supplying firm (+4 − +2.5 = +1.5, also exceeding the 'ideal choice'). The Ministry, or Minister and his officials, would be acting against the national interest, of course, for the outcome is not the ideal, but this inferior outcome, inferior that is from the point of view of the developing country, does arise often enough for us to have identified it previously as one common category, the conspiratorial choice. So the conspiratorial choice can be seen simply as the consequence of playing a one-person/many-person game with full knowledge under conditions permitting a co-operative solution with side-payments. A change in the rules is enough to vitiate the 'ideal choice'.

IMPLICATIONS FOR RESEARCH AND POLICY

Implications for Research

It may not seem strange that the game-theoretic analysis has revealed so much more about technological choice in developing countries than do other bodies of economic theory, but it must be admitted that the superior explanatory power of game-theory derives at least in part from the greater input of information. In order to be able to describe the alternatives to the 'stylised' outcome we had to provide a much more detailed description of the motives of the players, the structure and rules of the game, and the potential rewards to each player, than is necessary in, say, neoclassical microeconomics. Information is not free; it is expensive to gather, and the expected benefits to be gained should be compared to the expected costs of gathering it.

To anyone who believes that game-theory offers a useful approach to the choice of technology, priority in research should be given to the estimation of pay-offs. Formidable as this task is, the research involved would probably be no more demanding than was that involved in the first attempts at project appraisal in the developing countries. The benefits accruing from research into the application of game-theory to technological choice would be of the same nature: better understanding of the environment within which choice is made and of the range of feasible alternatives. Knowing what is possible is the initial step towards improving outcomes.

Pay-offs refer to specific outcomes to specific games, not general outcomes to general games. Research which yields estimates of pay-offs is therefore research into specific situations, into *this* project in *this* country at *this* time. Such studies would be subject to the same limitations as case studies generally are, and would be of value in proportion as they followed the same methodology, were conducted with the same honesty, skill and perseverance, and were numerous enough to enable generalisations to be made.

Implications for Policy Formulation

If research reveals that there are alternatives to the stylised outcomes, alternatives approximating more closely to the ideal, there are implications for government policy. Imagine that there was one alternative to the stylised outcome, more desirable from the point of view of the nation at large. It could then be argued that public policy should so be designed as to secure the more desirable outcome. The policy would be directed towards altering market structure and/or industrial organisation so as to guarantee that the participants selected strategies leading to the desired rather than the stylised outcome.

In game-theoretic terms, such policy can be seen either as eliminating the strategy pair (or n-tuple in n-person games) which leads to the stylised outcome, leaving pay-offs unchanged, or as altering pay-offs so as to make the most desirable alternative the one that the players will voluntarily choose. The two views of policy are not inconsistent but result from conceiving policy either as prohibiting certain actions or as encouraging other actions in their place. Prohibitions are merely the imposition of large negative pay-offs in the cells associated with what are declared to be economically undesirable strategies. Nevertheless, the second view of public policy, namely, that its purpose is to change the structure and rules of the game so as to encourage the selection of more desirable alternatives, is perhaps the more constructive; it does focus on the whole environment within which technological decisions are made and it does not lead one to underestimate the difficulty of designing efficacious public policies.

It might be interesting to observe the implications for the formation of public policy in the case of the stylised outcome analysed at some length on pp. 64–72 of this paper, the overly capital-intensive choice.

We recall from the analysis of the situation which led to the overly capital-intensive choice that this outcome maximised the total pay-offs to the two participants, the Ministry of Industry of the developing country and the foreign Aid Agency of the donor country. Beneficial this outcome also was to the economy of the developing country, but not so beneficial as the choice resulting in a less capital-intensive technology provided on generous terms and adopted promptly after speedy negotiations (see Figure 3.3). The objective towards whose attainment public policy in the developing country should be directed can be seen therefore as the choice of the latter, more beneficial outcome.

This much is obvious, but what are the further implications of the situation depicted in Figure 3.3? What can be seen besides the most beneficial outcome for the developing country (the north-west cell in the *right*-hand sub-matrix) and the most beneficial outcome for the two participants (the different outcome represented by the north-west cell in the *left*-hand side sub-matrix)? The first thing to be observed is that even if the government restricts the choice to the less capital-intensive technology (i.e., limits the feasible alternatives to those in the right-hand sub-matrix) there is another alternative that is more attractive to one of the participants (the Ministry), no less attractive to the other (the Aid Agency), but less beneficial to the developing country; this is the outcome in the south-east cell. It would be quite likely that the two participants, acting in their own interests, would choose to prolong negotiations and inflict severe terms on the aid recipient. The conclusion is that it would not be sufficient for the government to restrict the domain of choice to the less capital-intensive technology; necessary also would be additional public policies designed to prevent prolonging negotiations, so as to avoid the outcomes in the south-east and south-west cells, and giving a side-payment by one participant to another, so as to avoid the outcome in the north-east cell.

Three policy instruments, not one, are therefore needed to ensure that the choice made is that most beneficial to the developing country. One instrument would reject all alternatives to those involving the less capital-intensive technology, a second would assure that negotiations were conducted with dispatch, and a third would prevent side-payments. The implication for public policy is that the more alternatives there are to be eliminated from choice, the more instruments will have to be applied. Conceivably, as many instruments may have to be employed as there are alternatives more attractive to the participants than the outcome most beneficial to the country as a whole. In the case depicted in Figure 3.3 there were six alternatives more attractive to the participants than the optimum; in principle, therefore, no more than six instruments need to be applied. By a judicious choice of instruments, however, the number was reduced to three, one of those three (the restriction of the technology to the less capital-intensive one) eliminating the two alternatives in the right-hand sub-matrix more appealing to the participants, the second of those three (the avoidance of protracted negotiations) eliminating the two alternatives in the bottom row of the right hand sub-matrix, and the last of the three (the prevention of side-payments) eliminating the remaining alternative in the right-hand sub-matrix. In the face of these three policies, the participants would prefer the outcome most beneficial to the developing country.

Another implication of the same case is that the most beneficial outcome can usually be assured by more than one 'bundle' or 'mix' of policy instruments. Besides the combination of lower capital intensity/speedy negotiation/no side-payments, the identical outcome could be obtained by a combination of playing possible foreign donors off against each other/lower capital intensity/no side-payments.

It is worth noting that the prevention of side-payments is an instrument that appears in both policy 'bundles'; it may be an instrument for which there is no substitute. There may be other instruments which are necessary components of any 'bundle' too; one of the purposes of research would be to isolate any such policy instruments.

NOTES

1. For an application of the neoclassical theory of the firm to technological choice in developing countries, see David J.C. Forsyth (1980).
2. Examples of this extensive literature are André Gunder Frank (1969) and Stephen Hymer (1972).
3. The theory was invented by Neumann and Morgenstern (1947). The first major extensions were made by Luce and Raiffa (1957); subsequent developments are summarised in Bacharach (1976) and Friedman (1977). The most recent applications are discussed in two review articles; Kramer's (1977), written from the point of view of a political scientist, and Schotter and Schwödiauer's (1980), from the point of view of economists.
4. See Rapoport (1966).
5. In the 'prisoners' dilemma', as normally formulated, the pay-off matrix does not produce dominant strategies; the less attractive outcome results from each prisoner choosing that strategy which avoids the worst possible outcome.
6. In a rough and ready way, the 'stylised' outcome approximates the solution to co-operative games suggested by and named after Nash (see Bacharach, 1976, pp. 113–17).
7. This is the objective postulated for project appraisals carried out in developing countries; see Little and Mirrlees (1971), Dasgupta, Marglin and Sen (1972), and Hansen (1978).
8. See Forsyth (1980).
9. This, one of the most powerful theorems in game-theory, is due to Debreu and Scarf (1963) and Shubik (1959). It has been proved for trading games, and conjectured for games involving both production and trade (Hildebrandt and Kirman, 1975).
10. For a summary of the environment within which decisions on technological choice are made in the Republic of Korea, see Enos (1982).

REFERENCES

Bacharach, M.O.L. (1976) *Economics and the Theory of Games* (London: Macmillan).

Dasgupta, P., Marglin, S. and Sen, A. (1972) *Guidelines for Project Evaluation* (Vienna: UNIDO).

Debreu, G., and Scarf, H. (1963) 'A Limit Theorem on the Core of An Economy', *International Economic Review*.

Enos, J.L. (1982) 'The Adoption and Diffusion of Imported Techniques in South Korea', in Stewart, F. and James, J. (eds), *The Economics of New Technology in Developing Countries* (London: Frances Pinter).

Forsyth, D.J.C. (1980) 'Market Structures, Industrial Organisation and Technology', mimeographed World Employment Programme Research Working Paper; restricted (Geneva: ILO).

Frank, André Gunder (1969) *Capitalism and Underdevelopment in Latin America* (New York: Monthly Review Press).

Friedman, J.W. (1977) *Oligopoly and the Theory of Games* (Amsterdam: North Holland).

Hansen, B. (1978) *Guide to Practical Project Appraisal* (Vienna: UNIDO).

Hymer, Stephen (1972) 'The Multinational Corporation and the Law of Uneven Development', in Bhagwati, J. (ed.), *Economics and World Order from the 1970's to the 1990's* (New York: Free Press).

Hildebrandt, W. and Kirman, A.P. (1975) *Introduction to Equilibrium Analysis* (Amsterdam: North Holland).

Kramer, G.H., (1977) 'Theories of Political Processes', in Intrilligator M., (ed.) *Frontiers of Quantitative Economics*, Vol. IIIB (Amsterdam: North Holland).

Little, I.M.D. and Mirrlees, J.A. (1974) *Project Appraisal and Planning for Developing Countries* (London: Heinemann).

Luce, R.D. and Raiffa, H. (1957) *Games and Decisions* (New York: John Wiley).

Neumann, J. von, and Morgenstern, O. (1947) *Theory of Games and Economic Behaviour*, 2nd ed. (Princeton, NJ: Princeton University Press).

Rapoport, Anatol (1966) *Two Person Game Theory: The Essential Ideas* (Ann Arbor: University of Michigan Press).

Schotter, Andrew and Schwödiauer, G. (1980) 'Economics and Games Theory: A Survey', *Journal of Economic Literature*, June.

Schubik, M. (1959) *Strategy and Market Structure: Competition, Oligopoly and the Theory of Games* (New York: John Wiley).

4 On the Production of a National Technology

HENRY BRUTON

INTRODUCTION

The purpose of this paper may be stated as follows: technical change is essentially a problem-solving process whereby a society becomes increasingly better equipped to take advantage of its opportunities and to overcome the constraints inherent in its environment. The creation of an economy in which technical change of this sort is part of the routine of the economy is largely a matter of learning on the part of the members of the community. Few, if any, societies have resolved technological problems simply by importing technical knowledge or physical capital. While importing plays an important role, the real task is to create within the country the fundamental sources of technological development. Especially is it necessary for the national firm – private and public, small and large – to be involved in the knowledge accumulating process. More generally, the idea is that technological advance must occur in response to the conditions within the country's economic and social system rather than be imposed or simply made available from outside sources. In addition, the origins of the technological advance must apply to the economy in general rather than limited to scattered segments. A problem-solving economy is one characterised by a continuing flow of a myriad of small improvements throughout the economy which are easily and quickly applied. This essay studies some of the issues and ideas suggested by this way of thinking about a national technology.

The second section, following the Introduction, examines in some detail the knowledge accumulating process in a developing country. Attention is given to the fuller exploitation of currently underutilised knowledge and to the creation of new knowledge. The third section reviews briefly the particular roles of the multinational enterprise and the agricultural sector, while the fourth section considers a variety of policy issues.

KNOWLEDGE ACCUMULATION IN A DEVELOPING COUNTRY

The most explicit model of the knowledge accumulating process relates the output of new technical knowledge to the inputs of resources engaged in research. The allocation of resources to research constitutes an investment, and is justified on the basis of anticipated returns as are other forms of investment. Empirical evidence, chiefly from the United States,[1] lends some support to this way of thinking about the subject, but the findings can hardly be said to be conclusive, even for the United States. The major difficulty with this approach is that it does not explain, indeed does not seek to explain, the productivity growth that occurs over a wide range of economic activity where formal (i.e., measured) research activity does not take place. As already noted, a problem-solving economy is one characterised by a continuing flow of a large number of small improvements that affect just about every sector of the system. To understand how such a situation can exist, we need to probe in a somewhat different way from the straightforward profit-maximising, resource allocation model. It is convenient to begin this probing by considering a private firm functioning in a developing economy. After that, attention will be given to the other forms of economic organisation noted above. It should perhaps be noted here that formal training and education – usually called human capital formation – is relevant in many ways to the knowledge accumulating process. I shall have little to say on this subject, not because it is not important, but because there already exists a large literature on the subject, and it seems more useful to concentrate attention on other, equally pertinent matters.

The Firm of a Developing Country

The final step in the productivity increasing chain is taken within the producing unit, and therefore must involve calculations on the part of the producing unit's decision-makers as to the profitability of doing something differently from the existing routine. By definition, a change in the technology employed by a producing unit requires some change in the routine that had been in effect. Constantly changing production routines create the need for specific decisions by managers. Steady productivity growth means that a burden is placed on the firm's decision-makers that does not exist when the production process is unchanging over time.

In the light of these observations, the discussion of technological development in a developing country should begin with an analysis of the individual enterprise. To help create the appropriate perspective, it is useful to tell a story about a producing unit in such a country.

Demand and Search

The Need for Search. Consider then a firm that is functioning in an economy where any change over the past has been very slow. The unit has become very well meshed with the rest of the economic and social system within which it functions. The unit has a set of physical capital items, a labour force of given quality, a source of supply of produced inputs, a market, and a range of possible output of one or more commodities. The possible output levels along with the inputs imply a technology that is in use. The unit has been producing this way for some time. Not much has changed over the past except possibly that the unit has become more compatible with the environment in which it operates.

In this situation, the unit is earning a return that is looked upon as adequate to keep it functioning, but in most instances its decision-makers are not satisfied. They are aware in a vague way that 'things' could be better, possibly much better with resources already at hand. They know that their labour is idle a great deal of the time and that it could be used in a variety of ways to make the enterprise a more productive and rewarding place. They know that somehow with greater or different effort a better quality product and a more profitable output could be produced.

The decision-makers are also aware of the constraints that exist to prevent, or to make exceedingly difficult, any changes in the routine. Changes in labour routines are difficult and the results unpredictable. The nature of the supply curve of labour of various skills is not known or known only vaguely. Modifying the production routine to improve the quality of the product involves new instructions to labour and new organisation, and this requires in turn new data and new information. A change in routine may also involve changing the sources of inputs, and this may require establishing new credit and transportation arrangements and incurring risks not now present. The decision-makers find it easy to identify many sources of cost of change and of uncertainty of the outcome of changed routine. Indeed, they probably do not know – and do not know how to find out – the costs involved in such changes and they have equal difficulty in determining the increased returns associated with such changes even if they were as successful as expected. Available liquidity and access to funding are such that almost any risk must be avoided. Imperfect markets and market information also create constraints. For example, lack of access to cash can well be a constraint for many units even though in some real sense no such constraint should exist. And so the firm continues to go along pretty much as it has in the past.[2]

These characterisations of the domestic enterprise bear a family acquaintance to Harvey Leibenstein's notion of X-inefficiency.[3] Leibenstein makes a convincing case that few firms operate on their production frontier and cites data that show big differences in sectoral labour productivities among countries due to a variety of fairly easily corrected factors. Such differences he calls X-inefficiency. He explains such inefficiency mainly in terms of motivational factors. People do not work as hard 'as they could'. There is always some slack somewhere because contracts are incomplete, not all factors are marketed, the production function cannot be completely specified, and firms often collude to reduce uncertainty.

The argument here places the emphasis a bit differently. The firm is in some sense aware of its 'inefficiency', but is also aware of the costs and risks associated with change. Motivations are relevant, of course, but more fundamentally the question is one of incentives and inducements to search. In an important sense there is no well-defined production frontier for an individual enterprise. At a given time it is operating as it is operating for the reasons outlined above, and the technological objective is to induce a search for increased efficiency from whatever the existing position.[4]

It is a basic position of this paper that in order to create a steady flow of new knowledge over wide areas of the developing economy, the individual enterprise must be induced to generate a demand for such knowledge. And to induce such demand requires that forces exist, or be created, that induce search by the firm's decision-makers and that also induce them to accept the risks of introducing change.

The purpose of the preceding several paragraphs is to highlight both the importance and the difficulty of doing just this.

The Nature of Search. In the situation described above the firm's owners and managers see their production activity very much as a part of the total environment. The consequence is that they see no way in which to effect a change. Further, they see their position so much as a whole, so much in context, that they do not appreciate how to go about searching for the knowledge that they know they could profitably employ. It is seeing the producing unit as a unified, homogeneous whole that constitutes the chief barrier to innovate, the barrier even to search.

Anthropologist H.G. Barnett speaks of configurations.[5] In the context of the present discussion, the firm's operations and its place in the economy and the place of its managers in the society can be thought of as a configuration. Analysis then consists of breaking down this configuration in such a fashion that it is seen as a set of components or a set of configurations. Insight 'consists in discovering new potentialities that can be made manifest in other configurations . . . Insight means that a thing becomes different in some of its sensory manifestations'.[6] This sort of analysis and insight is necessary in order to provide a handle, a place to begin to change. More specifically, it is necessary in order to focus a search for means of improving the operation of the firm. As argued above, mere recognition that things are not very satisfactory and could be better is not a sufficient condition to initiate search, and certainly not enough to lead to the identification and introduction of a new technology.

It is, however, exceedingly difficult for a firm's managers to break down their configuration into components that facilitates identifying a starting point. Something then must happen from outside the firm's routine operations that focuses attention on one aspect that emerges as a specific, identified source of problems. This is an important point, the policy implications of which are explored later, but in general terms the argument is that an event that directs attention towards a specific aspect of the 'configuration', that calls attention to the advantages of doing something about that aspect, that is, creates a handle, thereby encourages search. The event must be such that it induces a search for ways to adapt, rather than ways to avoid adapting.[7] The units must not be induced to seek more protection, exemption from the new less favourable exchange rate, and so on.

It may help put the argument into a clearer focus to compare it with the common current practice of making more available; more funds, more foreign exchange, more advisors, more protection, more subsidies, more tax holidays, more scholarships, more etc. Such 'more' implies that the several sectors of the economy are pressing hard against an identified barrier that will be released by more of one or a few somethings. Where this pressure does not exist, then simply providing more is almost sure to lead to waste, to increased concentration of income, and other development defeating consequences. Compare also the above approach with that implied by the conventional theory in which the producing agent is assumed to await the appearance of a new technique or device from the research department or from a research organisation. Then the firm decides to buy or not to buy. In many cases this is more or less what happens in the developing country, particularly with funds made available at virtually zero cost to the individual who decides on the purchase. A not insignificant part of aid funds from all countries are frequently transferred in this fashion with the result that new machines and new technology are sent abroad only to remain idle and rust away. Similarly, a significant proportion of the existence of inappropriate technology found in developing countries has its origin in the fact that the using firms are passively waiting, in contrast to actively searching.

The Probability of Finding

If firms are induced to search, they expect to find and must find if continued, regular search is to be established. The next step, therefore, is to examine the probabilities of successful search, that is, the probability of finding a new technique that produces the increased welfare. There are two sides to this issue: on the one hand, there is the question of the source of supply of knowledge and the quantity of that supply; on the other hand, there is the question of the direction that search takes. This latter question is part of the more general question of inducing the production of new technical knowledge that will add to social welfare. The two sides are closely linked, and it would be helpful if both could be discussed simultaneously. However, it seems more useful to concentrate attention on one issue and then the other. We look first at the direction and form that search takes.[8]

In Figure 4.1 the firm is currently at position A. The axes identify plentiful resources and limitational resources. This language is used to remind ourselves that the issue is more complex than simply capital and labour. More generally, one could say that a movement

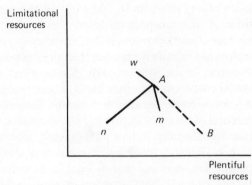

FIGURE 4.1

towards *m* would result in improved social welfare relative to *A*, while a move in the *w* direction would worsen social welfare. Is there anything in the system that will induce the firm's decision-maker to direct search towards *m* rather than *w*? Several factors may be identified as relevant.

First, the searcher has some clue, some feel, for the probability of success in searching in one direction rather than in another. The appraisals of these probabilities depend on a number of factors and, of course, vary from searcher to searcher. Presumably the greater the awareness of technical developments and possibilities in this general activity, the greater confidence the searcher will have in starting in one direction rather than in another. Awareness is in large part a consequence of the kind of information available to the searcher. Evident also is that awareness of possibilities in the neighbourhood of *A* is greater than that of possibilities distant from *A*. Thus if points *w* and *B* exist, the probability that the searcher knows that something may be possible in the *w* direction can be expected to be higher than that something in the *B* direction is hopeful.[9]

These same considerations also affect the cost of the search. Near search is less costly than far search, and search in a direction about which something is known is also expected (by the searcher) to be less costly than would be striking out on a completely random path.

Second, to move from *A* to another position may itself involve costs (i.e., down time, start up costs, new financing, etc.). These costs would be included in the comparison of expected costs at the prospective new position with those at the original position. Note should also be taken of the possibility that the extent of use of intermediate inputs may differ between the two positions.

In the absence of any effect on the demand for the product due to being at position A, the prospective differences in costs are all that matter. In this case if market prices reflect social costs, that is, if the plentiful resources are relatively cheaper than the limitational resources, then movement in the southeasterly direction will reduce both private and social costs. On the other hand, where prices to the firm do not identify social costs, the searcher may be induced to probe in the wrong direction in terms of improving social welfare. In this case distorted price signals may induce misdirected search. Thus the familiar pattern of interest subsidies, tax holidays, foreign aid for capital equipment, etc. may not merely lead to an initial position A that is socially non-optimal, but, more importantly, induce search in the wrong direction. The penalty for this latter wrong signal can be very harsh indeed. It may also be the case that if movement from A to w does in fact take place – because, say, the technological prospects are brighter in that direction – firms may then seek (and obtain) government policies that make w privately profitable, though socially unprofitable.

The appropriateness of search towards n is more complex. At n fewer of both plentiful and limitational resources are used. Whether operating at n does in fact contribute more to social welfare than does operating at A depends on the weights attached to the variables of the welfare function. If the weights attached to employment and to the reduction in severe poverty are very high relative to that attached to productivity growth, n may well be inferior to A. In any event, the profit-seeking firm will always find n preferable to A (assuming demand remains the same), and if low expectations of success in that direction do not offset the prospects of profits, the searcher will probe in that direction. To prevent movement to n will always require government subsidies or taxes of one sort or another.

Third, it is usually assumed that the quality of the product remains unchanged as an enterprise leaves A and hunts for another position. This is doubtless true for many products, but certainly not for all products. Even in agriculture, increased yields of fruits and vegetables may well be at the expense of flavour and nutritional value. For non-agricultural products, quality changes occur almost inevitably as the input mix changes.

If the search is induced by anticipated demand problems, then product prices may outweigh cost arguments. Thus the searcher may anticipate that at w the product price will be such that, despite the higher costs, the firm's profitability will be greater than at m. In this event, is the probing towards w socially acceptable? Again one must appeal to the welfare function. If the higher product price worsens the poverty and employment problem sufficiently to offset the increased (value) productivity then evidently w remains inferior to A and certainly to m.

Fourth, a final point is of great relevance. Suppose that the area south and east of A in Figure 4.1 represents improved social welfare relative to A while all other areas represent lower social welfare. Suppose further that market prices reflect social costs or are made to do so by a taxing and subsidy scheme. Then technologies found in other areas of the space will be recognised as less profitable than that employed at A, and hence not accepted. In these circumstances the new technologies that in fact appear will be increasingly appropriate, i.e., may use relatively more of the abundant resources or contribute to increased social welfare or do both. Presumably over time, searchers will recognise or seek to ascertain which direction to probe to find new knowledge that improves profit opportunities.

The Knowledge Producing Sectors

The preceding sections were concerned with creating a demand for new technology, that is, with inducing the firm to search and then inducing it to probe in the 'right' direction, i.e., the direction in which social welfare will be increased as a consequence of finding new knowledge. So now we have a domestic producer searching, and of course he must find. So the question now is the supply response to the search efforts.

There are two potential sources of supply that may be considered. In the first, the producer is directly involved in the creation of the new knowledge, in part by his own efforts, and in part by his active collaboration with others. The second is a somewhat more formal source, a capital goods sector, in which something like research and development takes place. For regular, cumulative technical change to take place both sources are necessary. It is convenient to begin with the more informal source in which the producer is an active agent in the knowledge accumulating process.

The Exploitation of Unutilised Knowledge

Earlier it was argued that the notion of a precise production possibility frontier was a doubtful instrument with which to illuminate economic reality. In the present context the chief difficulty is the notion of a 'given technology' that effectively limits the rate of output that is possible with available resources. (Of course, the notion of 'available resources' is also ambiguous, but it is not appropriate to consider this range of problems here.) 'Given technology' implies that all technical knowledge is defined by blueprints and all these blueprints are known to all producers. This, of course, cannot literally be the case in any country, and is an even less useful notion in a developing country. In the context of search, the appropriate given is that which is findable in 'prevailing circumstances'. Evidently this too is ambiguous, because one cannot ascertain what is findable until one has searched – and for how long? And what exactly are 'prevailing circumstances'? To help make the notion somewhat clearer, consider the following illustrations.

In several places in Africa one finds extensive junkyards.[10] Such junkyards are usually divided into numerous stalls. In an individual stall will be a specific collection of parts, pieces, and things which constitute the 'capital' of the junk dealers. In some instances this capital comes from elaborate, but derelict, aid projects. These dealers sell physical products, but more importantly they sell knowledge and technology. They sell ideas, and then supply an item to be used to implement that idea. The dealers are often extraordinarily clever in turning useless things into useful things in response to a specific request from a client. Clients are generally small-scale domestic enterprises and households. Evidently it makes no sense for such dealers to try to invent independently of the problems that their clients specify. The face-to-face meetings between the searcher and the potential rescuer is about as precise and continuous as it can be.[11]

These African junkyards are an example of a type of investment in knowledge accumulation that apparently has yielded high returns. Similar examples are becoming increasingly documented, although no inventory seems available.[12] They attest to the existence of an inventiveness on the part of certain members of the society, and to the existence of both the idea of search and the availability of a supply of a new knowledge. The knowledge thus created is invariably 'appropriate', and has emerged from within the society rather than being imposed on it from without. ('Appropriate' is, of course, a term that means different things to different people. In the present context, it refers to a new technology that is immediately compatible with the existing factor supply situation and the prevailing social environment.)

Such junkyards have a limited capacity to accumulate knowledge, but before examining the implications of that fact consider another example or two.

In Tanzania agriculture workshops were established in six villages with the help of outside experts.[13] Some tools were provided initially from beyond the village, and there was one formal course of instruction given by outsiders. Beyond that no outside help was provided, but there were periodic visits from the people of the Tanzania Agricultural Machinery Testing Unit at Arusha to suggest and encourage. The project's history is short, but observers emphasise two things of relevance to our story: the first refers to the importance of technology and new products (mainly agricultural implements and furniture) evolving in the context of their use. This not only helps to insure their appropriateness, but also encourages the flexibility and adaptability which in turn adds to factor substitutability. This latter is necessary to permit effective exploitation of locally available materials.

The second point is perhaps more important. The building of essential products that effectively meet the community's needs provides evidence that the community can do these things. The implements used no longer appear alien and necessarily store-bought or imported. The community learns to adapt and design and modify in important areas. As this is accomplished, then such attitudes and efforts will be transferred to other activities. Put differently, the experience suggests that new knowledge is in fact findable in the 'prevailing circumstances'. It does make sense, therefore, to try to induce a search mentality, because it is possible to find.

There is another side of this same point. The early introduction of new techniques created elsewhere or evolving from the production process almost always involve (as already noted) the acceptance of considerable risk. Hence the initiation in a given area is most likely to be done by the more affluent, the bolder, and the stronger members of the society. If these people have the capacity to import and to maintain alien techniques, then two important services are likely to go unperformed. The first is that the demand for locally created technical knowledge is severely penalised. Secondly, diffusion of new knowledge in village communities is largely through personal observation and informal contacts. Thus the less bold, the less affluent, the more risk averse must have an opportunity to see techniques available to them in action before they are likely to consider their use very seriously. By watching an affluent and respected member of the village try out new techniques, use new seeds, experiment with pest controls, buy locally made equipment, textiles, farm implements, tube wells, and so on, the community as a whole learns what is profitable to do and what is not profitable to do. The initial user therefore has a strategic role to play as a source of demand for new technology, and as a means by which that technology is diffused.

It is important to appreciate that imitation is an effective way of meeting uncertainty in all countries. It is especially effective in countries where information flows are whimsical and unreliable. Imitation is therefore a way of learning. It is a means of seeing what is findable. It also helps to explain the evolution of a system.

Consider a final closely related example. It refers to the contrast between an agricultural equipment workshop that fits in well with the local environment – factor supplies, social and political organisations, and psychological attitudes, etc. – and a new agricultural implements factory (e.g., a turn-key project built by a foreign aid donor).[14] The latter can supply a new type of (say) plough immediately. In the case of the workshop an opportunity for search and experiment is provided as is an opportunity for searcher and possible rescuer to explore together what is findable. Hence skills and complementary activities and infrastructure develop along with improved technology built into the agricultural machines. The technology evolves therefore as the other relevant parts of the system evolve. In the case of the new factory that turns out a new plough the aspect just mentioned is lost. The plough may well be more 'efficient' in some technical sense. It may also be so alien, so out of place, that foreign servicing is required or that it can be used only sporadically because of the absence of complementary activities and infrastructure. More importantly, it may bedazzle the prospective user to the extent that no learning, no questioning, no searching and finding is induced. Thus a static situation is created, not an evolving, responding process built on learning. The establishment of the latter is the heart of a social system in which technical change is an inherent part.

We are now at this point in the argument. Initiation of search for new knowledge by the user is the first step in creating a continuing flow of new technologies over broad areas of the economy. It is further argued that within 'prevailing circumstances' knowledge will be found if search is in fact initiated by the prospective user. The initiation of search by prospective users is the most effective way to ensure that the knowledge that is created is demanded, and therefore is appropriate. The argument that new technical knowledge can be found implies that there is within the community a great deal of findable knowledge, that is, knowledge that can be readily created when demand is exercised for a specific kind. No community has fully exploited all such knowledge, although of course some communities are nearer to doing so than are others.[15] The necessary first step in the pursuit of the objective of the creation of a society in which technology is constantly changing is the creating of search, of effort to find and exploit the knowledge that is in existence.

The task now is to examine the means by which this knowledge can be added to.

The Role of a Capital Goods Sector

The significant difference between the creation of knowledge and the finding of a specific piece of existing knowledge is found largely in the extent of the formalness of the search effort. The development of the principle of high yielding seeds illustrates the distinction very well. The general principle of such seeds was worked out in formal laboratories in a conscious research atmosphere. For this effort to pay off in higher yields around the world required search for the specific seed that would work in a given area. The main similarity between the two approaches is found largely in the necessity, in both instances, of those who are to respond to search to be in close and continuing touch with the searchers.

The not uncommon practice of building research institutes in order to accomplish this more formal task has generally proved unsuccessful. The commonly held view is expressed by Gustav Ranis: 'the developing world is strewn with scientific institutes and other expensive white elephants which contribute neither to science nor to technology'.[16] There are many explanations of such failure, but two are especially relevant in the present context. Such institutes are outside the real society; they are imposed on a society that is otherwise unprepared and unable to make use of them. These institutes are, in some sense, out of sequence with the development of the rest of the society, and hence cannot fit with and serve the community. The second difficulty is a consequence of the first. Most institutes have had little success in linking up with potential users of their output.[16a] The institutes seem as divorced from the reality they are expected to serve as are many universities.

Rather than research institutes the present analysis is built around the role of a domestic capital goods sector. This sector has two roles to perform: to create the new knowledge that adds to the findable knowledge and to screen and select the foreign technologies available for importing. Evidently the performance of both these roles requires considerable scientific (and economic) insight and understanding. The latter role will be discussed in the following section.

We will consider some broad observations, with a specific illustration or two, and then seek to identify a few general principles. The customary formulation of the comparative advantage argument rarely leads to the establishment of a capital goods sector in a developing country. Comparative costs are, of course, relevant, but must be formulated in such a way that the sources and processes of knowledge accumulation are recognised. The argument for the establishment of a capital goods industry is therefore linked to this latter issue. There are three general points: the first refers to the necessity for proximity between the capital goods sector and the using sector, the second to the kind of learning opportunities made available by a capital goods sector, and the third to the role of this sector in achieving further capital-saving changes.

The first point, the need for proximity, arises from a number of considerations, some of which have already been noted. The essential point is that the capital goods sector must be informed about the content of the searchings of using firms. More specifically, the capital goods sector must be thoroughly familiar with existing technical knowledge, and be concerned directly with the task of building from this knowledge. This sector must also be alive to the general environment within which the using firms function. Environment here refers to the kind of constraints that the producer faces, the input supply and demand situation, and the market structure. These constraints, as argued earlier, mean that the new knowledge created must be 'near' the present modes of production in the community. Near knowledge refers to new technologies that the using enterprise can recognise as potentially do-able and effective. It is something that will solve the problem identified by the using enterprise, but does not so violate the constraints that the latter firm simply gives up. Several surveys have found that the most important characteristic of a successful innovation was that it had its roots in an understanding of the needs of the user.[17]

Proximity is important for another reason. The full implications of a capital item, even a relatively simple one, cannot be communicated verbally, but can be appreciated only after extended use. Indeed, the capital item often must be modified after periods of use have revealed weaknesses or inadequacies. Thus close observation over time of new instruments and devices by their creators reveals information about their effectiveness and characteristics which may lead to modifications that add to their productivity. This is almost impossible unless the builder of the capital goods is more or less continuously available. This process has been emphasised by Nathan Rosenberg and identified by him as learning by using.[18]

The second argument for a domestic capital goods sector depends on it offering specific learning opportunities. At the outset of this essay technological development was defined as a problem-solving process whereby a society becomes increasingly able to exploit the opportunities available to it.[19] Emphasis is placed on 'problem-solving' and 'particular environment'. It does not mean – and we have seen over the past quarter century that it has not meant – simply being supplied with a larger output. The creation of the capacity to accomplish this process is a matter of learning by doing. What the capital goods sector can provide that cannot be provided elsewhere is this opportunity to learn. And this kind of capacity – skill is too narrow a word – is different from the capacity to operate a given machine or to perform an experiment in a laboratory. In seeking simply to import capital goods, the opportunity to acquire this capacity is effectively denied. An illustration may help.

Nathan Rosenberg emphasises the role of the machine tool industry in the development of the United States in the nineteenth century.[20] This activity played an important role in the solving of technological problems and in the diffusion of new technologies once they were created. The machine tool industry was especially important in the nineteenth century because of the dominant place that metals occupied. So people engaged in the use of machine tools became equipped to serve a great number and variety of industries. The machine tool industry in the United States by the 1950s was characterised by a large number of highly specialised, relatively small firms, each catering to a fairly limited and select clientele, and each working more or less in response to their customers' specific requests. As originators and as means of diffusion of new, applicable technical knowledge, these numerous small machine tool plants were extraordinarily important. In a very real way these firms did something else: they contributed to the creation of a category of people who not only had certain specific skills, but, more importantly, had a certain kind of attitude. This attitude may be characterised as confidence in the profitableness of search, the belief that technical and economic problems can be overcome. And this attitude in turn created confidence among their clients that it made sense for them to identify new profit opportunities, and then to ask the machine tool sector to create something that would enable these opportunities to be exploited.[21] Clock-making seems to have played a similar role as long ago as the thirteenth century. Cardwell states that clock-making gave rise to a very superior group of craftsmen who were able to contribute to a variety of activities. Much later clock and watchmakers played important roles in the development and operation of textile machinery.[22]

Developments of this kind seem to be taking place in some of the developing countries. In Colombia, for example, chemists and engineers trained for the nation's petrochemical industries have apparently begun to have some modest spillover effects on other sectors, mainly those closely related (fertilisers and paints). More importantly in the present context is the basis that domestic skills provide for a development of more productive and more widely dispersed technology. The skills of the silversmith can be immensely important in making or remodelling small machines, the skills of the potter are relevant to water distribution, those of the gunmaker to the working of metal.[23]

The third ground on which one may rest a case for a local or a nearby capital goods sector has to do with capital saving and labour saving, about which so much has been written and so little is known. Perhaps the most important form of capital-saving innovation is that produced by increased productivity in the domestic capital goods industry. If the quantity of real resources required to produce a piece of physical capital of unchanged quality declines, then a capital-saving innovation has occurred for the economic agents using that particular capital item. This situation would be reflected in an increase in productivity of capital in the using sector, although the new technology was developed in the capital goods sector. If the capital goods industry achieving the increased productivity is located within the country using the capital item, then that country enjoys an accumulation of technical knowledge that results in less capital per unit of output, irrespective of what happens to the price of the capital item. If, on the other hand, the increased productivity occurs in a capital goods sector abroad, then the price must fall in order for the importing country to experience a capital-saving technical change. It is, of course, one thing to achieve an increase in productivity in the capital producing sector, and quite another for the prices of imported capital items to fall.

One final observation on the capital goods industry question has to do with scale. It is difficult to avoid the idea that capital goods industries must necessarily be large-scale industries if acceptable costs are to be achieved. Evidently many capital goods industries in the world are in fact of great size. On the other hand, it is also evident that there are many producers who turn out machine tools of all kinds that are small-scale (i.e., small enough to thrive in a relatively small country without export) and by their very smallness are able to meet more effectively the needs of the community of producers that they serve. Engineering shops can be very small indeed, and can respond more readily to new opportunities than can very large producers committed heavily to a certain category of products.

Conclusions on the Capital Goods Sector

The roles that the capital goods sector must perform if it is to justify its establishment may be summarised:

(1) It must gain the capacity to respond to the identified problems of the capital-using sectors.
(2) It must acquire an understanding, a full knowledge of the constraints (and the firmness of the constraints) within which the using sector functions.
(3) It must provide the means of information flow between technical knowledge accumulation and prevailing search efforts by the knowledge users.
(4) It must contribute to the creation of a routine of problem identifying, of search, and of problem-solving that enables a society to experience sustained technological development.
(5) It must contribute to the creation of an economy where capital-saving productivity growth appears in a fairly routine way.

Evidently just any capital goods activity will not accomplish these objectives. In particular, it is important that the capital goods activities in the developing country must not seek simply to become copies of similar activities in the richer countries. The great danger is that efforts to establish a domestic capital goods sector will become just an unfortunate imitation, rather than a genuine indigenous activity. For example, the capital goods sector in India is frequently identified as quite successful, yet that sector in India has one of the most capital-intensive industrial sectors in the world.[24] More on this point in the section on policy.

MULTINATIONALS, PUBLIC COMPANIES, AND FARMS

The preceding discussion did not distinguish among categories of firms. That argument is broadly applicable to any domestic producing unit of a developing country. At the same time, it is useful to examine specific characteristics of three categories of firms – the multinational enterprise (MNE), the state-owned enterprise, and the agricultural enterprise – in so far as these characteristics are relevant to the technical knowledge creating process in the developing country.

The Role of the MNE in the Production of Appropriate Technology

It is useful to begin by identifying two general roles that the MNE may perform in the technology accumulating process. In the first instance, the MNE may bring in a different technology from that available in the country. Secondly, the MNE may contribute to the search and discovery efforts of domestic firms.

The main attention will be given to the second of the two roles, but a few comments on the first may be in order.

That the MNE can build a plant with a technology that is more productive – i.e., that yields a larger rate of output for conventionally measured inputs – than a domestic firm in a similar activity is hardly open to question. The questions that do arise are usually directed towards factor combinations employed, use of intermediate inputs, suitability of product, and possible effects on the balance of payments and income distribution. Considerable attention has been given to the capacity of the host country to affect by its set of policies the extent to which the MNE can (and will) bring a technology that has more or less favourable impacts in these several areas. Such an argument implies a 'shelf' of technologies from which the MNE chooses. Although the notion of a shelf of technologies raises a number of difficulties, it is fairly clear that government policies can have significant effects. At the same time, it must be recognised that most countries have had considerable difficulties in designing and implementing the 'right' policies. Few countries have, for example, established general policies with respect to foreign investments and refused to negotiate or bargain with potential investors. Many issues (fees for patent rights, transfer pricing questions, etc.) do not lend themselves to general regulations. The developing country does seem to have numerous difficulties in such negotiations.

These are complex issues, but they do permit the conclusion that the MNE can be controlled and its impact directed if a government can get its own policy in order and strengthen its negotiating capacity. To put it a bit differently, there is no inherent feature or characteristic of the MNE that makes it an inappropriate vehicle for transferring technical knowledge. There are, however, additional issues that make it necessary to worry about this conclusion, at least to some degree.

These issues concern the extent to which the MNE contributes to the creation of a domestic knowledge accumulating process. It is, as previous pages have argued, essential that such a domestic process be created if technical change is to become a genuine part of the routine operation of the system rather than an episodic event. If, for example, there are characteristics of the MNE that defeat or even dampen the creation of this metamorphosis, then it is necessary to reconsider whether appropriate policies and strength at the negotiating table are sufficient to ensure that MNEs earn their keep.

Several characteristics of the MNE that bear on this issue may be noted.[25]

The MNE's record on process innovations is not strong. Rather they are much more effective at creating and introducing new products. The available evidence also indicates that the MNEs do not have an outstanding record on the introduction of routine, run-of-the-mill innovations nor in the development and advancement of the scientific concepts that frequently are forerunners of an invention. Finally, the MNE depends very heavily on its home market to develop the new products. That new products especially, but all innovations, are closely related to the market conditions in which they are introduced has been demonstrated repeatedly. That MNEs perform along these lines (there are exceptions and qualifications, of course) is hardly surprising, given the usual formulations of the organisation and objectives of such enterprises.[26]

There are a number of implications of these characteristics relevant to the present argument. The concentration on the creation of new products by the MNE means that it tends to bring to the developing country a product with few linkages to the rest of the economy. (That the product often can be classified as a luxury is a consequence of inequality of income distribution within the developing country.) This means that inducements to create new supplying firms are minimal. In particular, the not uncommon practice in rich countries of engineers or other skilled workmen leaving the MNE to establish their own supplying firm is rarely observed in most developing countries. The kind of process that the discussion of the machine tool industry in the United States (see above) identifies is almost impossible to set in motion under such circumstances. Legal requirements that the MNE obtain some percentage of produced inputs within the developing country do not really have much, other than a distorting, effect. Also, in most instances, it appears that improved processes will yield greater advantage to the developing country than will a new product. The line between the two cannot be drawn unambiguously. What appears most useful at the present time to most developing countries is not access to detailed knowledge about a new product, but rather much more general concepts of design, of attributes of the properties of materials, testing procedures, and the like. These forms of knowledge do not lend themselves readily to transfer by a multinational enterprise. Finally, the fact that the MNE finds it so convenient to perform almost all its research and development activities in its home country means that there are few opportunities for the MNE to give support to, or even to come into contact with, any possible host country domestic research activity. The fact that the MNE does its R and D largely in response to market conditions in its home country is, of course, the main reason for the frequent unsuitability of product and technique for the developing country. There are exceptions to this general picture, but most such exceptions seem to result from public relations considerations rather than genuine commercial considerations.

These remarks suggest that the MNE is not likely to be an effective instrument for the creation of an indigenous technological capacity. Indeed, it is easy to find evidence that they may well harm any effort to do so. (There may be advantages of the MNE not connected with technology, but these are not within the terms of reference of the present study.) Some examples suggest that until there is an effective domestic technological change capacity, the foreign investor can do very little in this area. It is often noted that Japan gained greatly by borrowing and adapting technology from abroad. By the time it began to borrow, however, Japan had a strong domestic technology capacity and was well equipped to appraise, learn, accept risks and choose. Few contemporary developing countries are.

The Republic of Korea's more recent development suggests similar conclusions. Technology has been acquired from abroad, but very little direct foreign investment has been involved. The assimilation of the technology has been substantial and has been a Korean undertaking. Westphal and his associates emphasise that 'Korea's strategy to gain industrial competence has thus relied heavily on indigenous effort through various forms of learning by doing . . . and transactions at arm's length in the use of foreign resources'.[27] Similarly, in the industries that have experienced the most rapid development the technology has been of a general, non-proprietary type. In particular, it is useful to note that the Republic of Korea's big export success was not through multinationals or international subcontracting to MNEs. It is also evident that much of this rapid and 'appropriate' technological development did not represent a high level of sophistication in science and engineering. These are important lessons.

Technological Development Among Agricultural Units

The search and discovery of new knowledge in agriculture is especially relevant for three reasons: (i) this sector is such an important part of the economy of many developing countries; (ii) that new technology must in general be quite area-specific; and (iii) much of the technology employed in agriculture is developed in non-agricultural sectors.

In no country of the world do farmers themselves engage in formal research activity on any significant scale. It is, however, important to emphasise that farmers are perhaps the most strategic actors in this knowledge accumulating process. R.E. Evenson estimates that in the United States farmers spend as much as one-quarter of their time searching, screening, and testing new technology.[28] This surprisingly large estimate calls attention to the important role that the individual farmer plays, indeed must play, in any effort to establish a routine such that land yields and factor productivity increase as a matter of course.

There are two major assumptions on which the following argument builds. Accumulated empirical evidence of the last two decades has convinced virtually all observers that the farmer in all parts of the world responds strongly to price incentives. The evidence is especially convincing with respect to the choice of crops to produce. Perhaps the single most important, and elementary, condition for the constant emerging of new knowledge in agriculture is that the farmer be allowed to profit from his endeavours. It is, of course, common practice in many developing countries to penalise agriculture in the name of industrialisation or saving. At the same time it is equally important to recognise that making agriculture profitable is only a necessary, not a sufficient, condition for knowledge accumulation in agriculture. 'Sufficient' conditions vary widely from country to country. Evidently land tenure arrangements, access to credit, marketing facilities, etc., are relevant, and would have to be taken into account when examining agriculture in a particular country. The point here is that a pricing or tax policy that strongly penalises the profitability of agriculture will, under virtually all institutional arrangements, defeat the emergence of an effective technological development process.

The second assumption refers to the information and understanding enjoyed by the farmer. Numerous observers have noted that farmers have a great deal of insight and knowledge that is simply unavailable to anyone else, i.e., to extension workers, foreign experts, Ministry of Agriculture personnel, and so on. This knowledge is reflected not only in what farmers do at any given time, but in their capacity to identify their own needs, to appraise the suitability of new seeds, implements, practices, and to learn to exploit those new procedures that are deemed worthy of exploitation. Again one must add that this characteristic is not sufficient to carry the day. To understand how knowledge accumulation in agriculture occurs, the kind of knowledge the farmer has must be recognised and exploited, but it must also be supplemented. Any policy must build from and onto this knowledge. Again the point to note is the necessity to build from and onto rather than sweep away the old and impose some unidentified new.

Sectoral employment data show the proportion of the labour force in agriculture in northern North America, Europe, and Japan to be relatively small. In the United States the figure is 2 per cent, Australia 6 per cent, Japan 12 per cent, the United Kingdom 2 per cent, and so on. These data are often used to demonstrate the high productivity of labour in agriculture in these countries. More accurately, of course, they demonstrate the high productivity of all resources engaged in agricultural activities, and the latter include farm machinery manufacturers, seed and fertiliser producers, and a variety of other producers of intermediate inputs. These sources of intermediate inputs are closely and directly linked to the productivity of labour and land in raising crops. This close link between the intermediate input supplier and the farming community is another example of the importance of proximity of user and supplier in the knowledge accumulation process. The importance of proximity is enhanced by the area specificity of so much of agricultural knowledge. Although this location specificity is perhaps more unambiguous with respect to biological technology, it is important to appreciate that it also applies in significant degree to farm implements as well.

Finally, one should refer to the question of the sequence or order in which change occurs. Sequence is important in all sectors of the economy, but it seems especially relevant in agriculture. Thus multi-purpose dams are sometimes built long before the farming community is able to manage water, dig canals or obtain fertiliser. Similarly, the availability of high-yielding varieties of seeds has sometimes exacerbated income distribution, small farms/large farms conflicts and price fluctuations because they (the HYV) come along in the wrong sequence, that is, before other changes occurred in the system that made it possible to exploit, in the socially desired way, the advantages of the superior seeds. The sequence problem is especially evident in instances where an institution (research institutes, extension services, government organisations) is copied from abroad and set down in an environment in which existing institutions are incompatible. It is highly relevant, for example, that by the time tractors were becoming common on North American farms, the gasoline motor was already very familiar through experience with automobiles and small motors in general. Thus the move from mules and horses to tractors was a modest jump. For a farmer with little or no personal experience with gasoline motors to go overnight from using water buffalo to a tractor is, in most instances, an impossible leap.

What do these several points add up to in so far as the task of creating an indigenous knowledge accumulating capacity in agriculture in modern developing countries is concerned? The contention here is that the starting point must be the making of agriculture production activities very profitable. Suggested ways of doing this are discussed in the following section. Profitable production opportunities in agriculture have several consequences. In the first place, they help to induce the farmer to search or begin to search for ways to exploit these opportunities. They make him alive to the rewards of search, and help him accept the risks inherent in a new technique. They create a situation where the farmer's knowledge, previously emphasised, can be brought into play. As noted the full exploitation of this knowledge is a necessary building block towards the ultimate objective. The full extent of this knowledge is not known, indeed cannot be known (even to the farmer), until the farmer is pushed to the outermost limits of his capacity.

Secondly, the increased demand by the farmers will create some, possibly a substantial, inducement to invest by other sectors linked to agriculture. Though empirical support is not available, the frequent failure of activities that produce intermediate goods – implements especially – to expand favourably is surely in many instances simply due to the absence of a demand for their services. The blacksmiths, small metal shops, the junkyards noted earlier, all offer bases from which a response may occur. In the 1880s, for example, there were around 800 distinct models of ploughs available in the northern United States, all produced by small firms. Most were ploughs designed to meet specific needs of the localities in which they existed. Generally such firms came into being in consequence of opportunities created by a profitable agriculture.[29] To repeat, policies that squeeze agriculture in developing countries tend to defeat the evolving of associated activities about which there is available, but unfound and unexploited, knowledge. The importance of climate, soil, rainfall, etc. in farming combined with the deeply established social and cultural circumstances of agricultural communities tends to make extremely difficult, perhaps impossible, the simple application of technologies imported from abroad. Indeed it may be argued that one reason why so many countries have penalised or neglected agriculture arises from their determination to imitate the West, and imitating the West in manufacturing was less obviously nonsense than doing so in agriculture.[30]

The preceding processes, even where they work well, cannot be depended on to carry the whole burden of agricultural development. Publicly supported experiment stations are necessary as are more formal research arrangements as the process evolves and gains strength. Such should not, indeed cannot be, rushed. What needs to be widely established is the idea of inventing in response to observed opportunities, and the establishment of any such idea is a time-consuming process. While large-scale foreign training programmes or imports of foreign experts have often resulted in considerable harm to the knowledge accumulation effort in agriculture, it may be that with a more selective approach these kinds of programmes can be designed in a way that will help. It is, of course, crucial that whatever use is made of foreigners, such use must be part of the process just described, not in place of it. The most important function that includes foreign participation is the maintenance of a continuing relationship between the domestic agricultural community and those organisations where very basic research – generally chemical and biological research – is being done. The Green Revolution was largely the working out of a method of inventing. Some parts of this method may be transferable as may be some aspects of certain findings. These can supplement a local effort, but in no sense replace it. A bit more on policy matters is considered in the last section.

The Role of the State-Owned Enterprise

The role that state-owned or controlled enterprises can play in the development of an indigenous technological capacity has not been explored as much as has the role of the multinational. Consequently generalisations are more difficult to identify and establish. In principle one may argue simply that a state-owned enterprise is no different from one that is privately owned. It is easy to convince oneself that a government has enough to do without concerning itself with the ownership and management of firms. A government that can resist the temptation to build, or seek to build, the artefacts of the West, and that can therefore apply its energies to establishing a general economic environment in which the kind of search and selection process outlined in previous sections will emerge, may, in most instances, contribute more to the development of technological capacity than it will in the role of enterprise owner and manager.

It is the purpose of the following section to address directly this role of the government.

TOWARDS A KNOWLEDGE ACCUMULATING POLICY

The general theme of the preceding discussion may be summarised as follows: the fundamental objective is the emergence of an indigenous technological capacity. The accomplishment of this objective depends very much on the mobilisation of existing and currently unused indigenous knowledge. Some of this knowledge is not used because the fact of its existence, or its findability, is not recognised. Incentives must therefore come into being that induce search for and diffusion of this presently unexploited knowledge. The accomplishment of the objective depends also, of course, on the evolving of the internal capacity to create genuinely new knowledge that is consistent with the social and economic characteristics of the country. To do this seems necessarily to call for some kind of links with those countries where the knowledge accumulating process is already well established. Perhaps the most difficult and risky aspect of the whole process is that of finding an appropriate means by which links with more developed countries may be established.

Some policy issues have already been briefly discussed. In this section a more direct confrontation with the policy question is undertaken. As a beginning, some generalisations on government policy are noted, and this is followed by some more specific observations.

General Role of Government

An explicitly formulated national technology policy is helpful for a variety of reasons. The most obvious reason is that it forces the government to address the issue directly. Such a policy must be more than a series of truistic statements. A government that is explicitly pursuing a development policy should be able to answer with considerable clarity the question: what is your policy with respect to technological development?

The most specific role that a government is in a position to perform is to support search efforts by its procurement policies. The government here appears as a user, a demander of new knowledge, and, as emphasised repeatedly above, users accumulate special knowledge and understanding about the products that they employ. Communication between the government as a user, or prospective user, and the producers of knowledge enables the government to pass that knowledge on to producers, but, more concretely, it is able to offer direct incentives that will induce search. The point here is that the government is seeking to achieve its own purposes, and, in so doing, it can evaluate and select new knowledge and new technologies as an informed buyer. Governments do not do very well as the agency to determine what new technologies, or category of technologies, are to be pushed. As a buyer, it can exercise power by selecting and buying.

One final point along this same theme may be noted. The government may undertake to establish procedures and mechanisms by which all prospective users of new knowledge play a major role in what goes on in the capital goods sector, i.e., in the sector where new knowledge is being generated. Exact methods for doing this will depend very much on local institutions and routines, but some explicit government action may be in order.

Some More Specific Policy Approaches

Specific policies must necessarily emerge from the environment of the country to which they apply. It is not possible, therefore, to prepare a detailed blueprint of a set of policies directly applicable to any one country independent of that country's particular position. The following observations are intended as directions or opportunities emerging from the arguments of the previous pages.

(i) In most developing countries one observes prices of factors and products that are distorted to a (in a number of cases, large) degree. These distortions are a consequence of many things, some of which are deep-seated institutional matters and some are largely policy matters. Many countries have explicitly followed policies of making capital cheaper (via tax holidays, investment credits, etc.) than it would have been in the absence of such policies. The usual objective of such policies is to encourage capital investment. Economists generally condemn such practices, but governments seem reluctant to abandon them. One reason for this reluctance is the belief that some form of explicit encouragement to investment is necessary, and alternative policies are not readily available.

The discussions in the second and third sections do in fact lead to alternative forms of subsidies. It is easy to design practicable incentive systems that reward increased employment and increased productivity in a firm. Such incentives can also be designed that induce a search for ways to produce intermediate goods domestically, and are a means of highlighting the advantages and profitability of searching and changing production routines in a particular direction. Such instruments apply more readily to non-agricultural enterprises than they do to farms. And they require that fairly detailed records, or at least the capacity to check the firm's records, be of some thoroughness. It will therefore reach only the more organised, tightly run enterprises. This observation, of course, applies equally to the tax holiday arrangements presently widely used. (More detail on the financing of such policies is provided below.)

(ii) To create additional inducements for the small-scale farmer and the less well-organised manufacturer poses greater problems of both design and implementation. In many situations specific incentives may be less urgent than for the larger, more organised sector because the prices the latter faces are likely to be more distorted and misleading. At the same time, it is in the smaller units that improved production would, in many instances, yield the highest social returns. So direct attack is very much in order. On the basis of the technology accumulation process worked out in previous sections, it is possible to make some general policy suggestions.

(*a*) In a number of countries the government purchases a variety of agricultural products directly from the farmer. Where this practice is in operation, the government has some machinery in place to influence decisions of the farmers and to convey information to them and obtain information from them. Similarly, the government often is the sole source of supply of intermediate goods, fertilisers, pesticides, seeds, etc.

In these instances a pricing strategy can be devised that will signal to the farmer the advantages of proceeding in a particular way. The idea is to confront the farmer with rising marginal costs of these inputs, yet to make it highly profitable to him to increase output. To accomplish the latter, prices paid for the agriculture produce should *rise* as yield per land unit increases. Hence a strong inducement is created to find ways to increase yield without increasing the use of intermediate goods.

(*b*) Mechanisation of agriculture in developing countries is enormously complex, and the inclination is to discourage it in labour surplus economies. Appropriate mechanisation can, however, add to land's productivity, and thereby increase the demand for labour. To relate price received positively to yield per land unit will also help to induce search for machines whose use makes land yields rise. In general, if mechanisation results in output per unit of land rising more than output per labour unit, we can expect labour use per land unit to rise. Given the incentive systems described in the previous paragraph, search would be induced in that direction (assuming labour to be the plentiful resource).

(*c*) In countries where the government does not purchase agricultural commodities as a general rule, reaching small-scale farmers with fiscal measures would require setting up other mechanisms. (Large-scale farms would presumably be covered by the tax arrangements described above.) Even for smaller farms where produce is marketed directly by the farmer, it may be administratively feasible to establish an arrangement whereby such farmers are subsidised as their yields increase and purchased intermediate inputs reduced. In this event the government would pay directly to the farmer some amount based on the percentage increase in yields and the percentage decrease in purchased inputs. Such a process would require considerable monitoring and record-keeping by a government agency. The task would be cumbersome, but not complex. If this kind of a scheme were instituted, even small farmers would be expected to search for ways to increase yields. This search for implements would bear mainly on local, small-scale 'engineering' shops, the indigenous capital goods sector, from which some response can be anticipated.

(iii) The preceding discussion placed major emphasis on seeking to induce economic agents to do what they would not otherwise do because of distorted signals, lack of knowledge, risk aversion or whatever. In contrast to most approaches these proposed measures do not make certain things cheaper; they make certain things more costly, while offering a reward for the finding of appropriate new means to increase output. This kind of measure requires funds, and much depends on how governments get the money to finance subsidies. It also involves administrative capacity. Several observations are pertinent.

(*a*) A proposal that could be very effective has to do with foreign assistance. Foreign aid is generally in the form of projects to provide physical capital or training of people. But aid could be used to finance the kind of subsidies and incentive systems discussed above. Evidently there are many details that would have to be worked out, but one can conclude rather readily that aid used in this form would be much more productive than a great deal of the aid that has been provided in the past. Aid has seemed to have helped the most where policies were effective in directing efforts along lines consistent with the preceding arguments (e.g., Republic of Korea).

(*b*) Conventional foreign aid projects that build dams and highways perhaps have a net favourable impact, but in many instances they do result in bringing in new technologies that dampen interest in the search for domestic solutions. Aid also confuses the picture of determining correct market signals. Even where aid provides research tools and training programmes, it may result in making the long-run solution to the technological question more difficult by diverting attention from the domestic demand and supply scene to activities that are in no way responsive to the expressed needs of the society.

Still one cannot argue the eschewing of aid by a developing country. So the argument must be that stated above, namely, seek to use aid chiefly as a means to establishing effective policy instruments. Once that is done the dangers of aid are much less, and the possibilities of using aid to raise social welfare will be greatly increased.

(iv) On the international side of things the general direction in which policy should move is fairly clear, but to give full flesh to these directions is a matter that depends very much on the particular characteristics of the specific country. The following points are generally relevant to all developing countries.

(*a*) Almost all new activities in such countries need protection and competition. Too much protection can discourage effort, too little can create a sense of hopelessness – and eliminate incentives to search. Protection and pressure are both necessary, but an atmosphere that creates the feeling among economic agents that the answer to the appearance of any threat or problem is more protection or more aid is sure to eliminate most inducements to search. For a significant part of the economy, the foreign trade sector can be used as a means of providing the 'right' amount of competition. It is by creating opportunities to export, that pressures and incentives are produced that generate active search for ways to exploit these opportunities. Similarly, by making imports a threat, but not allowing them to overrun the fragile domestic activity, an incentive to search is created.

(*b*) Along with tariffs the most potent specific instrument of policy in this regard is the exchange rate. An 'undervaluation' (by standard criteria) of domestic currency relative to foreign currency is a powerful inducement in a great variety of respects. 'Undervaluation' provides a strong inducement to export, and adds to the protection against imports. The main point is not primarily to earn foreign exchange. Opportunities to export create incentives to improve quality, to reduce costs, and to exploit domestic resources. The undervaluation also makes imports more costly, further inducing searching domestically. It is this set of incentives and pressures that constitute the fundamental impact of protection and exchange rate policy. Perhaps it is not too strong a statement to say that the general practice of many developing countries to maintain an 'overvaluation' of their currency along with the great panoply of import substitution policies has so effectively reduced competition or its threat that few economic agents found much reason to search. It is a principal conclusion of this study that the most effective link between domestic producers and foreign enterprises of any kind is created by strong inducements to export. In particular, it is emphasised that technology can be transferred in this way most securely.

(*c*) Evidently the common practice of admitting capital goods at zero or very low tariffs is contrary to the arguments worked out earlier. Capital goods activities should receive about the same rate of protection as other activities. This point is now generally recognised, and also generally violated.

(v) In the earlier discussion on the role of the MNE, attention was directed to the necessity of the existence of an internal technological capacity in order to realise longer term advantages from its presence. Attention, therefore, has been directed to the extent to which the MNE contributes, or can be induced to contribute, to the creation of this technological capacity. A few more specific points on the MNE are possible.

(*a*) It is helpful to remind ourselves that in the period before the Second World War, most of the international capital movements were portfolio rather than direct investment. This was especially true of British capital movements in the century before the First World War, and meant that the recipients of this capital had considerably more freedom than is now generally the case. Such capital transfers went largely to countries recently settled by European immigrants.[31] Thus these capital movements did produce a transfer of technology, but it was accompanied by freedom of use by a community not unlike the community from which it came.

The Japanese experience from the 1880s was quite different but is equally relevant. The Japanese, as noted earlier, were well equipped with their own technological capacity and were a remarkably homogeneous society, and were therefore enabled to select from abroad the technical knowledge that they could employ in their own environment. There was, of course, no direct investment by foreign companies in Japan. It was the prior existence of this capacity to choose and adapt that made the Japanese so effective in their technology importing efforts.

(*b*) There are examples of MNEs performing as agents of technology transfer, but they do not seem to constitute a suitable general instrument for such a task. Technical knowledge will be, and should be, transferred among nations. The question refers to the means by which that transfer is to be effected. With minimal foreign ownership of enterprises but extremely strong export incentives, one may argue that technology will be transferred in a fairly satisfactory fashion. More general problems created by foreign ownership and foreign presence will also be reduced. Such a policy guide does not preclude a MNE presence for one or two specific opportunities or tasks, but minimises its role significantly.

(*c*) The preceding argument rests on the assumption that the developing country can avoid gross distortions in its economy. One frequently reads or hears that modern development requires capital-intensive activities that use complex technology, and hence MNEs are necessary and technological dependence is inevitable. This paper has been at pains to dispute this conclusion. Considerable technological dependency does in fact exist, but it exists because developing countries have chosen a route that defeats the emergence of their own technological capacity.[32]

A Final Word

One could continue discussing policy matters indefinitely. The policy area included here has been the general fiscal activities of the government. Other areas are also of great importance. Elementary education is especially relevant in the kind of argument worked out here. Further attention to the role of formal research institutes could be informative, as could further investigation of insurance policies to spread the costs of risks and uncertainty. And on and on.

The last word is about the same as the first word. An indigenous technological capacity enables a nation to respond to opportunities, to create opportunities. It also enables the society to pursue new paths that it identifies and chooses and that emerge from the inherent nature of the society. That is after all what development is about.

NOTES

1. Mansfield (1968) provides a convenient summary of this evidence.
2. The situation is not unlike that which Professor Evsey Domar describes as prevailing in the USSR.

   The problem in the USSR has been the absence of a built-in mechanism for the introduction of inventions into industry. If anything, the system has developed an automatic device for the rejection of such inventions, or at least long delay in their acceptance.

   The learning-by-doing process enables an enterprise manager to make a given activity and to reduce costs. Thus an old product is both easy and profitable to produce since its price, once established, remains fixed for a long time. A new product or process involves all sorts of difficulties and uncertainties, and the price for it is fixed at the outset but a small profit (Domar, 1969, p. 45).

   See also Berliner (1976).
3. Leibenstein (1966) has written extensively on this topic.
4. This point and its development in later parts of this paper employ many of the notions and arguments of Nelson and Winter (1982).
5. See the opening chapters of Barnett (1953).
6. Ibid., pp. 186 and 411.

7. In his study of development in Japan, Lockwood (1954) emphasises the 'cumulative importance of myriads of relatively simple improvements in technology which do not depart radically from tradition or require large units of new investments' (p. 198). He also points out that the large, modern, capital-intensive factory played a role, 'but the real substance of Japanese economic growth, however, is found in the more modest types of improvements which are more easily and pervasively adopted, more economical in cost, and often more productive of immediate return and income' (Ibid., p. 199).

8. The argument developed in the following pages makes use of ideas found in a variety of sources, especially Nelson and Winter (1982), Binswanger *et al.* (1978), Bruton (1977). Binswanger *et al.* employ diagrams similar to that used here, but in a different way.

9. Findings for the United States show that a large proportion (almost three-quarters) of R and D activities are aimed at very limited advances in the particular of research. See Mansfield *et al.* (1982), chapter 10.

10. I am grateful to my former colleague, Robert Schneider, for calling my attention to this illustration.

11. In a major survey of the success and failure of innovations in Great Britain, Christopher Freeman found that the single measure which discriminated most clearly between success and failure was 'user need understood'. Freeman emphasises that the notion of 'user need understood' does not mean market research, but rather that the new product or service that constituted the innovation was designed to meet an understood need of future users. This characteristic overshadowed such things as size of firm, size of R and D department, numbers of qualified scientists, etc. These results are reported in Freeman (1973).

12. See OECD (1977 a and b) and Stewart (1974) for some examples.

13. Based on the report by MacPherson and Jackson (1975).

14. Based on an example in OECD (1975).

15. Attention may again be called to Leibenstein (1966) and Nelson and Winter (1982).

16. See Ranis's paper in Beranek and Ranis (1978), p. 22. This paper is an extremely useful survey of several aspects of the knowledge accumulating process.

16a. Editor's note: For some empirical evidence, see pp. 168, 224 and 235 in the present work.

17. See Freeman (1973) and note 11 above.

18. See Rosenberg (1982a), chapter 6.

19. Compare Rosenberg (1976), p. 523.

20. Rosenberg (1974), chapter 1.

21. Many development policies concentrate energies on creating industries, in contrast to agriculture and mining. Arguments supporting these policies usually appeal to income and price elasticities and the terms of trade. A more important argument follows from the point in the text, namely, much of the labour in the developing countries is engaged in activities that offer little opportunity or scope for learning, for seeking out and responding to new situations, for achieving an understanding of the importance of adaptation and change. Hence the occasions for learning are very limited.

22. See Cardwell (1972).
23. See Bhalla (1975) and OECD (1976).
24. Little (1982) argues in this way, and notes that it greatly dampens the success of this sector in contributing to India's long-run development.
25. The discussion in this paragraph relies heavily on Vernon (1977). See also Caves (1969) and Dunning (1974).
26. Caves (1969) is a good review of the general organisational aspects and objectives of the MNEs. This book also has extensive references.
27. Larry Westphal and his colleagues have studied the Korean experience in some detail. The quotes are from Westphal, Rhee and Purcell (1981), pp. 65–6. Also Little ((1982), pp. 224–5) has some helpful observations.
28. Evenson's very interesting paper is in Nelson (1982).
29. These data are from Evenson in Nelson (1982).
30. This point is made by Solo (1966) and others.
31. Rosenberg (1982) emphasises these points and develops a number of their implications.
32. This is a major theme of Little (1982, pp. 240–8).

REFERENCES

Barnett, H.G. (1953) *Innovation: The Basis of Cultural Change* (New York).
Beranek, William, Jr. and Ranis, Gustav (eds) (1978) *Science, Technology, and Economic Development* (New York: Praeger).
Berliner, J.S. (1976) *The Innovation Decision in Soviet Industry* (Cambridge and London: M.I.T. Press).
Bhalla, A.S. (ed.) (1975) *Technology and Employment in Industry* (Geneva: International Labour Office).
Binswanger, Hans P. *et al.* (1978) *Induced Innovation* (Baltimore: Johns Hopkins Press).
Bruton, Henry J. (1977) 'A Note on the Transfer of Technology', *Economic Development and Cultural Change*, Supplement.
Cardwell, D.S.L. (1972) *Technology, Science and History* (London).
Caves, Richard E. (1969) *Multinational Enterprise and Economic Analysis* (Cambridge University Press).
Domar, Evsey D. (1969) 'Theory on Innovations' (Discussion), *American Economic Review*, May.
Dunning, J.H. (1974) *Economic Analysis and the Multinational Enterprise* (London: Allen and Unwin).
Freeman, C. (1973) 'A Study of Success and Failure in Industrial Innovation', in B.R. Williams (1973).
Leibenstein, Harvey (1966) 'Allocative vs. X-efficiency', *American Economic Review*, June.
Little, Ian M.D. (1982) *Economic Development* (New York: Basic Books).
Lockwood, William W. (1954) *The Economic Development of Japan* (Princeton, N.J: Princeton University Press).
MacPherson, George and Jackson, Dudley (1975) 'Village Technology for Rural Development', *International Labour Review*, January.

Mansfield, Edwin (1968) *The Economics of Technological Change* (New York: W. Norton).

Mansfield, Edwin *et al.* (1971) *Research and Innovation in the Modern Corporation* (New York: W. Norton).

Mansfield, Edwin *et al.* (1982) *Technology Transfer, Productivity and Economic Policy* (New York: W. Norton).

Nelson, Richard R. (1982) *Government and Technical Progress* (New York: Pergamon Press).

Nelson, Richard R. and Winter, Sidney G. (1982) *An Evolutionary Theory of Economic Change* (Cambridge, Mass: Harvard University Press).

OECD (1975) *Transfer of Technology for Small Industries* (Paris).

OECD (1976) *Appropriate Technology, Problems and Promises* (Paris).

OECD (1977a) *Transfer of Technology by Multinational Corporations* (Paris).

OECD (1977b) *Science and Technology in the People's Republic of China* (Paris).

Rosenberg, Nathan (1974) *Perspectives on Technology* (New York: Cambridge University Press).

Rosenberg, Nathan (1976) 'On Technological Expectations', *Economic Journal*, September.

Rosenberg, Nathan (1982) 'The International Transfer of Industrial Technology: Past and Present', in OECD *North/South Technology Transfer*; *Analytical Studies* (Paris).

Rosenberg, Nathan (1982a) *Inside the Black Box: Technology and Economics* (Cambridge University Press).

Solo, Robert (1966) 'The Capacity to Assimilate an Advanced Technology', *American Economic Review*, May.

Stewart, Frances (1974) 'Technology and Employment in LDCs', *World Development*, March.

Vernon, Raymond (1977) *Storm Over the Multinationals* (Cambridge, Mass: Harvard University Press).

Westphal, Larry, Rhee, Yung W., and Purcell, Gary (1981) *Korean Industrial Competence: Where It Came From*, World Bank Staff Working Paper, No. 469.

Williams, B.R. (1973) *Science and Technology in Economic Growth* (London: Macmillan).

5 The Role of Appropriate Technology in a Redistributive Development Strategy

JEFFREY JAMES[1]

In the 1950s and 1960s when growth of GNP was widely considered to be the primary goal of developing countries, technology was discussed mainly in terms of its contribution to this goal. Sen (1968) and others, for example, argued that capital-intensive techniques would lead to faster growth because they are associated with a higher rate of investible surplus than relatively labour-intensive techniques.

During the 1970s, however, it became apparent that GNP growth alone would not necessarily solve the problems of mass poverty and unemployment. Indeed, despite unprecedented high growth rates between 1950 and 1975 in many developing countries, there has been both a rapid rise in the numbers of unemployed and an increase in the numbers who live in poverty.[2]

It also became evident that one of the reasons for the dualistic pattern of growth was the use in the modern sector of most developing countries of advanced capital-intensive technologies imported from the West. In particular, 'the resource requirements of advanced country technology has involved high levels of investment per man, concentrating LDCs' scarce investment resources on a small proportion of the population with the consequence of un- and underemployment and low productivity for the rest of the population'.[3]

Attention consequently shifted to the need for redistributive growth strategies and to the advocacy of alternative, more appropriate technologies for contributing towards a reduction in inequality and the alleviation of mass poverty. In much of the subsequent discussion, however, there was a tendency to simply *assume* that these technologies should form part of a redistributive strategy. Rarely were the merits and demerits of policies for appropriate technology compared to those of alternative redistributive instruments. As a result, the possibility that there may be circumstances in which the latter are preferable to the former has seldom been contemplated. (Put another way, few authors questioned the logicality of the view that technology should form part of the solution to inequality, simply because it contributed to the problem.) The aim of this paper, accordingly, is to assess the role that policies for appropriate technology might play in different kinds of redistributive development strategies. For this it is necessary (i) to examine the extent to which these policies are *in principle* capable of reaching the heterogeneous target groups among the poor, (ii) to examine the range of alternative policies that are available to policy-makers, (iii) to assess the merits and demerits of these alternatives in relation to appropriate technology, and (iv) to identify complementarities between policies for appropriate technology and other policy instruments.

We begin with an examination of the first of these issues.

THE ECONOMIC RELATIONSHIP BETWEEN APPROPRIATE TECHNOLOGY AND POVERTY

As noted above, one of the major consequences of inappropriate technology is economic and technological dualism – the co-existence of a high productivity modern sector employing a small proportion of the labour force, with a low productivity sector in which the majority of the population are employed and underemployed.

Appropriate technology is often defined, conversely, with reference to its contribution to *reducing* this type of dualism.[4] Because the reduction can take place at both ends of the technological continuum, – on traditional as well as on modern technology – the effects of appropriate technology on poverty and inequality operate in two different and separate ways. The division of the economy into two sectors – traditional and modern – is, of course, oversimplification. Nevertheless, it provides a convenient basis on which to approach the main issues.

The purpose of modifying modern sector technology is to increase employment opportunities, particularly those of an unskilled kind.

The technologies that are capable of achieving this increase may already exist (but for various reasons remain unchosen) or they may need to be created through adaptation/generation. From the point of view of this dimension of appropriate technology policy, it is clear that only those among the poor who were formerly unemployed are capable of benefiting. But there may be some among the poor unemployed or underemployed who, as Sen[5] has pointed out, 'do not look for work elsewhere and could not be persuaded to take up wage employment elsewhere even if it were offered to them'. That is, it cannot be assumed that all those among the poor who are unemployed (underemployed) can be helped through the creation of additional wage employment opportunities.

Modification of traditional technology, however, will have a more diverse impact on the various groups living in poverty. For one purpose of improving this technology is to improve productivity and raise the incomes of those who possess productive assets (such as small farmers, artisans in the urban informal sector, and so on). In addition, depending on the system of wage determination, improved technology may benefit those workers whose productivity has increased.[6] The first distinction that needs to be made here is between wage and non-wage modes of production. In the latter case, where the family system predominates and incomes are determined by the average product, an increase in productivity will benefit all the family members to an equal extent. On the other hand, where wage employment prevails, it is the individual workers whose wages may or may not increase following a rise in productivity. In classical-type models, for example, productivity gains are appropriated by the owners of the means of production. The constancy of the real wage in the 'capitalist' sector faced by a perfectly elastic supply of labour, is a feature of the well-known Lewis model of development. In the neoclassical model, on the other hand, wages are determined by the (marginal) productivity of labour (though imperfections in factor or product markets mean that labour may receive less than its marginal product even according to this theory).

Whether or not direct employment increases as a result of improving traditional technologies depends on whether the decline in unit labour requirements associated with the rise in labour productivity is fully offset by an expansion of output. In part, this depends on the price elasticity of demand for the output. If demand is elastic, an expansion of output leads to only a small reduction in price. Under competitive conditions, for example, in which each producer faces a

TABLE 5.1 *The direct impact of appropriate technology*

Effect of appropriate technology	Group among the poor Unemployed	Working poor	Owners of productive assets
Rise in wages	×	+	× or –
Rise in employment	+ or ×*	×	×
Fall in employment	×	–	×
Rise in profits or earnings	×	×	+

+ = gain from change
× = position unchanged
– = lose from change

* This depends, as noted above, on whether or not the unemployed (under-employed) are willing to accept new wage-employment opportunities.

highly elastic demand curve, there is no discouragement from this point of view to an expansion of output associated with an increase in productivity. On the other hand, where demand is inelastic – as in the case of some agricultural products – increased labour productivity could reduce the demand for labour. In this event, the gains to some of those living in poverty from the application of appropriate technology would have to be weighed against the losses to those who formerly were employed (though the extended family sharing system that prevails in most developing countries may, to some extent, act as an automatic equalising influence on individual gains and losses).

Improved traditional technology is capable, therefore, of impinging on the welfare of numerous low-income occupational groups but the outcome need not always have an unambiguously favourable impact on inequality. Table 5.1 sets out these various groups and the probable direct impact on their incomes of the various dimensions of policies for appropriate technology.

It is, of course, entirely possible that *none* of the diverse groups comprising the poor benefit from improved technology. This will occur, for example, if the improved technologies are adopted mainly or only by those with incomes above the poverty line and if wages and employment of those in poverty both fail to increase. Something like this appears to have occurred with the new technology of the Green Revolution in many countries. At the other extreme is the case in which all the poverty groups benefit but none of those living above the poverty line are able to do so. Of course, in reality most cases will fall somewhere in between these two extremes, with the final impact on inequality depending upon the particular combination of those who gain, those who lose and those whose position is left unchanged.

The disaggregated way of looking at the impact of appropriate technologies on poverty and inequality that is embodied in Table 5.1 differs somewhat from the view of many advocates of appropriate technology who tend to see poverty in more homogeneous terms and who, in consequence, tend to adhere somewhat uncritically to the maxim that 'small is egalitarian'.

The above discussion has been concerned with only the direct impact of policies to secure appropriate technologies. Table 5.1 lists only these direct effects. Yet, in addition, there will be a series of indirect effects which may either reinforce or counteract the initial redistributive effects. To the extent that appropriate techniques make fuller use of local materials and resources, the linkage effects are likely to reinforce the initial effects. About the direction of other types of indirect effects, there is, however, little that can be said *a priori*. Nevertheless, it is important to recognise that they may often be substantial. Adelman and Robinson,[7] for example, conclude their simulation analysis of redistributive policies in South Korea with the warning that 'immediate effects rarely reflect the overall impact of a policy intervention, and the partial-equilibrium solution is seldom quantitatively close to the final-equilibrium solution. Indeed, in many of our experiments the ultimate effect is opposite to the initial impact or partial-equilibrium effect'. What they found was that there are powerful interactions between the various sectors of the economy that affect the responses to policy interventions. They found, for instance, that changes in the terms of trade for wage-goods, consequent upon a rise in the incomes of the poor, have a powerful secondary impact on the initial redistribution.[8] In the following section, where we examine the direct impact of technology on the structure of poverty in different countries, the limitations of the partial-equilibrium approach need, therefore, to be borne in mind.

Finally, it is important to note that although we shall be concerned in the paper only with *process* technology, new *products* may also have a redistributive impact. Consider for example, the situation portrayed in Figure 5.1.[9]

Suppose the consumer is initially at R. New good Z (assumed for ease of exposition to be indivisible) is now introduced enabling the individual to reach S on the higher indifference curve IC_2. If, however, good Z' had been developed instead (so that P could be

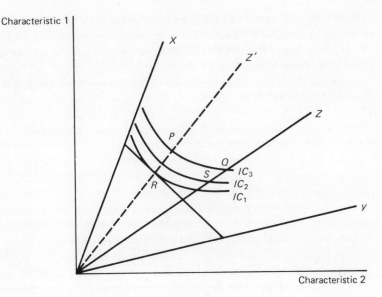

FIGURE 5.1

reached) *or* if Z had been available at a lower price (so that Q could be reached) the individual would have been even better off on indifference curve IC_3.

Because almost all new goods are developed in and for rich countries, there is a systematic inegalitarian bias in the world innovation system. Deliberate efforts are required to offset this bias by the design of goods embodying characteristics in proportions that match the tastes of the low-income majority in the Third World (that is, through the design of so-called appropriate products). In this way the level of welfare associated with *given* low incomes can be increased.

APPROPRIATE TECHNOLOGY AND THE HETEROGENEITY OF POVERTY

In the previous section we showed that while appropriate technology is capable of benefiting the majority of groups comprising the poor, it could, depending on the circumstances, have a differential impact on these groups, benefiting some of the poor, leaving the position of others unchanged and harming the position of yet others. In such a case, where appropriate technology has a non-uniform impact on the poor, its usefulness as a redistributive policy therefore depends on the relative proportions of the different groups in poverty in the particular country concerned.

Consider the situation in the sample of four countries contained in Table 5.2.

TABLE 5.2 *Occupational composition of the poor (percentages)*

	Employer	Self-employed	Employee	Housewife	Unemployed
Malaysia, 1970 (poorest 49%)	0.5	51.8	41.8	2.3	3.6

	Employer	Self-employed	Employee, private sector	Employee, public sector	Sharecropper
Brazil, 1960 (poorest 31%)	0.5	51.0	37.0	3.0	8.0

	Employer	Self-employed	Salary earner	Worker
Chile, 1968 (poorest 46%)	0.0	24.0	5.0	71.0

	Employer	Self-employed	Employee	Unemployed	Other
Trinidad and Tobago,* 1975/6 (poorest 42%)	0.4	15.1	25.6	18.6	40.3

* The table on which these figures are based classifies households by monthly income and occupation of head of household.

SOURCE Griffin and James (1981), p. 26.

The table illustrates, first, the heterogeneity of poverty within each of the four countries. That is, it shows how diverse are the occupational categories of those classified as living in poverty. It also shows how the size of the groups in poverty varies substantially from one country to another.[10] In Trinidad and Tobago a relatively high proportion of those classified as poor are unemployed, whereas in Malaysia poverty clearly has little to do with unemployment. Indeed, in Malaysia, as in Brazil, over half the poor are self-employed. In Chile, in contrast, poverty is mainly a problem of the 'working poor' since well over half the poor are wage earners.

The importance of this variation in the composition of poverty between countries is immediate, for it means that a given policy of appropriate technology will have a quite different impact depending on where it is applied. For example, an improvement in the productivity of traditional technology which is not also associated with an increase in employment (because of demand inelasticity) will do little to solve the poverty problem of Trinidad and Tobago (which is importantly associated with unemployment). Moreover, it is very likely that the degree of inequality will worsen. Similarly, unless the working poor of Chile are able to share in the productivity gains associated with an improvement in traditional technologies through higher wages, the result will be only a fairly slight impact on poverty and perhaps even an increase in inequality. The extent to which the Chilean working poor would, in fact, be able to share in any productivity gains depends mainly on their relative strength *vis-à-vis* employers – in turn, a function of the degree of unionisation, the role of government, etc. And in Malaysia and Brazil it is mainly the self-employed who need the benefit of improved technology if poverty is to be substantially reduced.

In our assessment of the role of appropriate technology as a redistributive instrument we have thus far dealt with the question of its impact on heterogeneous poverty groups. We have argued that it may or may not be capable of reaching a large proportion of those individuals classified as living in poverty. But even in situations where (the economic relationships are such that) technology policy *is* able to effect the desired reduction in poverty among the target groups, one has still to ask whether there are not alternative policy instruments that can achieve the same result more effectively. The answer to this depends, of course, not only on the range of alternatives available but also on the meaning that is given to the concept of 'effectiveness' according to which these alternatives are to be compared with appropriate technology. It is to a discussion of these issues that we now turn.

THE RANGE OF POLICY OPTIONS

In principle, there is a wide variety of policy options available to governments wishing to effect a reduction in the degree of inequality in the distribution of income. But the range that is *actually* available will vary a great deal according to the political situations that prevail in particular countries at particular times.

In most Third World countries of the mixed-economy type, governments owe their allegiance to powerful groups among the rural and urban élite and measures that run counter to the interests of these groups – particularly those that do so on a large scale – are usually not politically feasible. The most that can generally be expected from these governments, therefore, is a series of minor distributive reforms. In the context of these reforms, certain policies – especially those which involve a sharp change in asset ownership in favour of the poor – will generally not be available to policy-makers.

There are, however, political configurations in which governments are able to pursue a more radical redistributive strategy, in which the choice of instruments is much less restricted. It is the role of policies for appropriate technology in these, albeit somewhat rare, kinds of circumstances to which we first turn our attention. Thereafter, the role of appropriate technology in the more conventional, reformist type of context will be briefly considered.

Radical Strategies

In a radical redistributive strategy the emphasis, mainly for political reasons, is on the speed (and perhaps also the visibility) with which the reduction in inequality can be effected. This emphasis on rapidity is due to the fact that a radical strategy is normally associated with a political revolution, the dynamics of which generally require that the redistributive programme be pushed through at maximum speed. When discussing the problem of redistribution in a socialist context, Lange,[11] for example, in an often-quoted remark, has noted that

> A socialist government has to decide to carry out its socialisation at one stroke, or to give it up altogether. The very coming into power of such a government must cause a financial panic and economic collapse. Therefore, the socialist government must either guarantee the immunity of private property and private enterprise in order to enable the capitalist economy to function normally, in doing which it gives up its socialist aims, or it must go through resolutely with its socialisation programme at maximum speed. Any hesitation, any vacillation and indecision would provoke the inevitable economic catastrophe.

In at least the early phases of the programme, policies for appropriate technology, because they are relatively slow acting, are consequently unlikely to have much appeal. That such policies take considerable time to implement is obvious where the technologies have to be created. But even if this is not the case, and the technologies already exist, there are nevertheless often considerable lags in the search for, and installation of, them. (These lags are not merely due to the inevitable delays that occur even in rich countries, but result additionally from the particular difficulties caused by the absence of promotional efforts for the sale of most forms of appropriate technology.)[12] Moreover, policies for appropriate technology tend to lack the visibility or directness that often and understandably seems to be required of policies to redistribute incomes in the aftermath of a political revolution. For both of these reasons, therefore, the scarce administrative resources of policy-makers will tend to be concentrated on alternative policies, among the most important of which are land reform and policies to increase employment through expansionary policies on the demand side (such as wage increases) designed to make use of the considerable amount of excess capacity which usually exists in LDCs. Together, these policies may often be able to bring about a quite substantial reduction of poverty and inequality within a short period of time.

The effectiveness of land reform in the reduction of poverty varies considerably from one region to another. Most countries of Latin America, for example, have an abundance of land which is usually also subject to extremely concentrated ownership. Combined with the fact that average incomes in Latin American countries are relatively high (compared to Asia and Africa), this means that 'the worst of the rural poverty problem . . . can be overcome by an effective, implemented land reform'.[13] In the case of most African and Asian countries, in contrast, where fertile land is scarce relative to population and average incomes are low, even a radical land reform would leave large numbers in absolute poverty. In relation to India, for instance, Minhas (1970) showed that a transfer of some 43 million acres of land from the largest land-holding group would merely reduce the proportion of the population living in absolute poverty from about 40 per cent to about 33 per cent.

In the type of situation which we are considering, the success of policies to expand employment through an expansion of aggregate demand depends on the extent of excess capacity. In Cuba, the existence of resources that had remained un- and under-utilised prior to 1959 allowed for increases in output and employment between 1959 and 1961.[14] During the Allende regime in Chile, the availability of substantial amounts of excess capacity was made a cornerstone of the radical redistributive strategy. In 1970, when Allende came to power, idle industrial capacity and unemployment were among the highest ones recorded in that decade.[15] Members of Allende's Popular Unity government therefore expected that an increase in aggregate demand would lead to an increase of both production and employment, an expectation that proved, at least initially, to be justified.

The expansion of employment through the use of excess capacity can be regarded as a form of technology policy. In Figure 5.2 the increase in employment from L to L_1 that is brought about by the increased capacity utilisation of technique X (from K to K_1) is equivalent to that which would result from the adoption of the more labour-intensive technique Y.

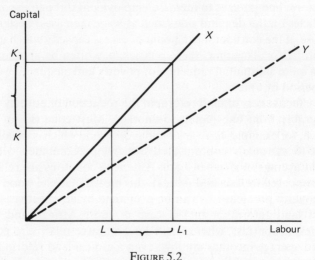

FIGURE 5.2

Once full-capacity utilisation of existing techniques has been reached, however, unless these techniques are appropriate, further attempts to redistribute income via employment creation will encounter increasingly serious problems of inefficiency. The reason is that in the absence of a 'capital-saving efficient technique to provide jobs for all willing and able to work'[16] the reduction in inequality will be brought about with a considerable wastage of resources in the form of unproductive labour. The essence of the problem can be brought out in the oversimplified form of Figure 5.3.

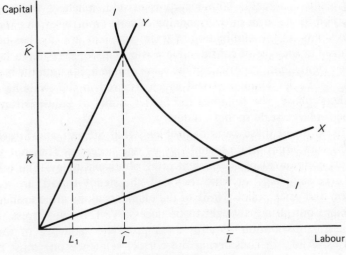

FIGURE 5.3

II is a familiar neoclassical isoquant showing efficient combinations of capital and labour that are available to produce a given level of output of a single good. $\hat{K}$ and $\hat{L}$ represent the amounts of the two factors in a notional capital-abundant, labour-scarce type of economy. $\bar{K}$ and $\bar{L}$ are assumed to represent the factor-endowment in a typical developing country. Full employment of the two factors will be achieved in both countries when, respectively, techniques Y and X are used. Following Eckaus (1973), let us assume that only one technique, Y, is actually available, i.e., that capital and labour have to be used in fixed proportions. Then, in the developing country, even assuming full capacity utilisation, unemployment will be equal to $\bar{L}L_1$. And given the redistributive aims of the government, with the implication that these workers cannot simply be left unemployed, they will have either to be given the dole or, what amounts to much the same thing, to be put to work unproductively (i.e., in the form of overmanning).

The Cuban experience illustrates this type of problem rather clearly. Thus, the new factories installed in Cuba after the Revolution were highly capital intensive and provided few employment opportunities. 'Cuban planners discovered in the early 1960s what the Chinese had realised a few years before, that thirty years of technological progress made the Soviet miracle of rapid unemployment absorption through industrialisation difficult to repeat.'[17] Partly as a result, a sizeable labour surplus existed, much of which was absorbed in the form of overmanning of state enterprises. As Mesa-Lago[18] puts it, 'the elimination of unemployment in Cuba has been achieved at the cost of considerable waste of national resources'. To some extent China experienced the same problem, for until the break with the Soviet Union in 1960 which ushered in the 'walking-on-two-legs' phase, she followed the Soviet model of industrialisation based on large-scale modern industry.

It is to avoid these sorts of problems that Stewart and Streeten (1976) view appropriate technology as one necessary (but not sufficient) component of an efficient redistributive strategy. The other two necessary 'legs' of what they call 'the steady tripod' are asset reform and price policy. None of the elements alone are considered sufficient but all are thought to be necessary. They also stress that efficient redistribution requires an appropriate sequence of these policies. And, 'In considering the correct sequence one must pay attention to the pressures which reform on one front will have in inducing or preventing reform on the other two, and the likely impact of the one reform taken by itself.'[19] This consideration provides additional support for the argument adduced earlier that policies for appropriate technology will tend to follow rather than precede asset redistribution. For without an initial impetus to equality, the markets for the type of low-income products normally associated with appropriate technology simply do not exist. One example of the complementarity between these two policies relates to an earlier period in China. Thus, 'an essential aspect of China's textile technology has been to produce large quantities of standardised cotton outfits, rather than higher quality textiles in a variety of colours and styles. But this required a market for such cottons'.[20]

So far, we have analysed the role of appropriate technology in a large-scale redistributive strategy that is rapidly implemented. As noted above, however, the far more common type of situation is one in which the redistributive options available to policy-makers are much more limited and tend to be confined to gradual reductions in inequality.

Gradualist Strategies

For the 'reformongering' type of strategy, the requirements of redistributive instruments are quite different from those that are important in the radical strategy. Whereas in the latter, the emphasis, at least in the early stages of the strategy, is on the speed and visibility with which the redistribution can be brought about, the reformonger is concerned essentially with the *feasibility* of alternative policies. (Feasibility, as noted above, is not really an issue in a radical strategy where almost all policies are likely to possess this quality.) And it is less overtly redistributive policies that are more likely to serve this goal than the direct and highly visible measures that are sought after in a radical redistribution. What appeared to be disadvantages of policies for appropriate technology in the earlier discussion – namely, their relatively slow speed and their somewhat indirect mode of redistribution – may therefore turn out to be more like advantages in the context of a gradualist strategy.

Whether technology policies are more feasible than traditional gradualist instruments (most notably fiscal measures), however, is a question about which generalisation is difficult, and which, partly for this reason, has tended to be ignored.[21] A major difficulty here lies in the fact that appropriate technology comprises not one but a variety of *different* policies and outcomes, each of which may have different implications in terms of their feasibility. In the first section of the paper, we observed that the gainers from appropriate technology policies could vary substantially according to the particular circumstances. In some cases the gains will be concentrated among a single group in poverty, and in other cases they will tend to be much more widely diffused. If the strength of the various groups in poverty varies substantially, then the feasibility of policies for appropriate technology which have different outcomes – because they depend partly on the relative strength of gainers and losers[22] – will also vary. Moreover, there are many different ways of attempting to secure the adoption of more appropriate technologies. And each of these different methods is likely to be associated with a different outcome in terms of the composition of gainers and losers. For example, a policy to change relative factor prices will generally have a different matrix of costs and benefits from a policy to restrict the production of certain goods or to divert resources to the generation of small-scale technologies for village use. As a result, the political feasibility of appropriate technology will depend upon exactly which set of policies is to be used in its implementation.

The feasibility of policies for appropriate technology will also vary according to their impact on important goals of policy other than income redistribution. In general, policies which, in addition to their redistributive impact, have a favourable impact on major goals of government, such as economic growth, appear to have a greater prospect of implementation than those with a negative (or neutral) impact. There are some technology policies, for example, that would, through a shift to alternative techniques, permit an increase in real national income *as well as* a rise in employment. These possibilities arise in circumstances where existing choices are technically inefficient, that is, where they use more of all resources to produce a unit of output than available alternatives. A recent study of public enterprises in Tanzania found that there was a high degree of technical inefficiency among them – indeed, no fewer than 78 per cent of the parastatals studied used more of both capital and labour to produce a unit of output than other firms in their industries.[23] Where, as in this example, the choice of more appropriate techniques is associated with an improvement in resource allocation it is possible, in principle, for *all* groups to benefit. That is, it is theoretically possible for those who gain from the change (e.g., the formerly unemployed) to compensate those who lose (e.g., suppliers of capital-intensive machinery) and still remain better off. (In the language of welfare economics, a potential Pareto improvement can thereby be converted into an actual improvement.) As Clive Bell[24] has described the possibility, in the context of the feasibility of redistributive policies in general: 'In practice, of course, there is almost bound to be some group or other which would lose as a result – unless they were to receive compensation from the rest. The trick here is to choose interventions which would cause the poor to gain, and then to devise specific means whereby rich – and politically powerful – losers could be *actually* compensated, while leaving the poor better off than before.'

In reality, however, compensation may be difficult to effect, particularly if losers comprise a heterogeneous and geographically dispersed group. For not only may those who lose sometimes be difficult or impossible to identify, but policies that benefit *only* these groups may also be hard to devise. As a result of these difficulties, even policies for appropriate technology that improve resource allocation (and which consequently satisfy the Kaldor–Hicks criterion) will frequently leave at least some people worse off. And, depending on the relative political strength of these individuals, policies which save resources may be no more feasible than those which use up resources (e.g., a switch to technically inefficient techniques) or which leave their total unchanged (e.g., a switch from one efficient technique to another).

Even from this brief discussion, it is evident that we need to know a great deal more about the factors that determine the political feasibility of particular policies in specific circumstances before any realistic assessment of the role of appropriate technology in a gradualist strategy is possible.[25]

SUMMARY AND CONCLUSIONS

The point of departure of this paper was the view that while the notion of appropriate technology evolved out of a dissatisfaction with growing poverty and inequality in the Third World, only a relatively small amount of attention has been paid to the role that it might actually play in a development strategy designed to reduce or eliminate these problems.

In order to formulate such an assessment, we began by noting that appropriate technology may or may not be capable of reaching a large proportion of those individuals classified as living in poverty. Then, to further the assessment, a distinction was drawn between two types of redistributive development strategies. In the first of these – the radical strategy – the aim is to achieve a large-scale redistribution of income in a short period of time. We argued that in the initial stages of this strategy, technology policy is more likely to take the form of an expansion in the capacity of techniques actually in use, rather than a shift towards more labour-intensive alternatives. Thereafter, in the longer run, an expansion in the markets for appropriate products will permit, and the exhaustion of idle capacity will require, the pursuit of more conventional policies for appropriate technology. In the second type of redistributive strategy – which we termed gradualist – the emphasis is on the feasibility of policy instruments rather than the speed with which they can effect a given redistribution of income. The political feasibility of alternative policies for appropriate technology, and hence their claim for inclusion in this type of strategy, was thought to depend mainly upon the matrix of costs and benefits to which they give rise, as well as their impact on important goals of policy other than income redistribution.

All of these conclusions, it should be stressed, are more or less tentative; they are intended to stimulate further research in this underexplored area rather than to offer firm conclusions to policymakers who are concerned to pursue redistributive development strategies.[26]

NOTES

1. I am grateful to Ajit Bhalla and Susumu Watanabe for comments on an earlier draft of this paper.
2. See Morawetz (1977).
3. See Stewart (1979), p.165.
4. See Stewart (1974).
5. See Sen (1975), p.39.
6. See Griffin and James (1981).
7. See Adelman and Robinson (1978), p.103.
8. Similar findings were reported in the BACHUE series of economic–demographic models. See Hopkins, Rogers and Wéry (1976).
9. This analysis, and the figure on which it is based, appear in James (1983). See also James and Stewart (1981).
10. To some degree these variations may be due to conceptual/statistical differences. For example, concepts used in Trinidad may raise the unemployment rate relative to those used in Malaysia.
11. See Lange (1964), p.124.
12. One is referring here to sales by domestic capital-goods producers, agents of used-equipment dealers and representatives of producers of appropriate capital goods in developed countries. See Pack (1982).
13. See Bell and Duloy (1974), p.102.
14. See Ritter (1974).
15. See Griffith-Jones (1978).
16. See Stewart and Streeten (1976), p.401.
17. See Mesa-Lago (1972), pp. 53–4.
18. Ibid., p.64.
19. See Stewart and Streeten (1976), p.404.
20. Stewart (1974), p.30.
21. See Cooper (1973).
22. See Stewart's paper in this volume.
23. See Perkins (1983).
24. See Bell (1974), p.59.
25. See Ilchman and Uphoff's (1974) critique of the ILO Kenya Mission based on this point.
26. This is also the conclusion reached by Stewart, in her essay in this volume.

REFERENCES

Adelman, I. and Robinson, S. (1978) *Income Distribution Policy in Developing Countries* (Oxford University Press).
Bell, C.L.G. (1974) 'The Political Framework', in H. Chenery *et al.*, *Redistribution with Growth* (Oxford University Press).
Bell, C.L.G. and Duloy, J.H. (1974) 'Formulating a Strategy', in H. Chenery *et al.*, *Redistribution with Growth* (Oxford University Press).
Cooper, C. (1973) 'Choice of Techniques and Technical Change as Problems in Political Economy', *International Social Science Journal*, vol. 25, no. 3.

Eckaus, R.S. (1973) 'The Factor-Proportions Problem in Underdeveloped Areas', in A.N. Agarwala and S.P. Singh (eds), *The Economics of Underdevelopment* (Oxford University Press).

Griffin, K. and James, J. (1981) *The Transition of Egalitarian Development: Economic Policies for Structural Change in the Third World* (London: Macmillan).

Griffith-Jones, S. (1978) 'A Critical Evaluation of Popular Unity's Short-Term and Financial Policy', *World Development*, July–August.

Hopkins, M.J.D., Rodgers, G.B. and Wéry, R. (1976) 'Evaluating a Basic-Needs Strategy and Population Policies: The Bachue Approach', *International Labour Review*, November–December.

Ilchman, W. and Uphoff N. (1974) 'Beyond the Economics of Labor-Intensive Development: Politics and Administration', *Public Policy*, Spring.

James, J. (1983) *Consumer Choice in the Third World* (London: Macmillan).

James, J. and Stewart, F. (1981) 'New Products: A Discussion of the Welfare Effects of the Introduction of New Products in Developing Countries', *Oxford Economic Papers*, March.

Lange, O. (1964): "On the Economic Theory of Socialism" in O. Lange and F.M., Taylor, *On the Economic Theory of Socialism* (New York: McGraw-Hill).

Mesa-Lago, C. (1972) *The Labor Force, Employment, Unemployment and Underemployment* (Beverly Hills, Calif: Sage).

Minhas, B.S. (1970) 'Rural Poverty, Land Redistribution and Economic Strategy', *Indian Economic Review*, vol. 5.

Morawetz, D. (1977) *Twenty-Five Years of Economic Development* (Baltimore: Johns Hopkins University Press).

Pack, H. (1982) 'Aggregate Implications of Factor Substitution in Industrial Processes', *Journal of Development Economics*, August.

Perkins, F.C. (1983) 'Technology Choice, Industrialisation and Development Experiences in Tanzania', *Journal of Development Studies*, January.

Ritter, A. (1974) *The Economic Development of Revolutionary Cuba* (New York: Praeger).

Sen, A.K. (1968) *Choice of Techniques,* 3rd ed. (Oxford: Blackwell).

Sen, A.K. (1975) *Employment, Technology and Development* (Oxford: Clarendon).

Stewart, F. (1974) 'Technology and Employment in LDCs', *World Development*, March.

Stewart, F. (1979) 'International Mechanisms for Appropriate Technology', in A.S. Bhalla (ed.) *Towards Global Action for Appropriate Technology* (Oxford: Pergamon Press).

Stewart, F. and Streeten, P. (1976) 'New Strategies for Development: Poverty, Income Distribution, and Growth', *Oxford Economic Papers*, November.

Part II
Empirical Studies

6 Government Policy, Market Structure and Choice of Technology in Egypt

DAVID J.C. FORSYTH

INTRODUCTION

Intensive study of issues concerning technology in recent years has made it clear that the range of factors conditioning technology choice and development in LDCs is wide, and that the precise definition of what constitutes 'appropriateness' is itself complex, depending crucially on the specific macroeconomic and microeconomic context. A consequence of this broadening of the scope of the technology debate is that an ever-widening range of government activities can be seen to be relevant to, and, indeed, to comprise, technology policy, though the majority of such activities are not viewed in this light by the authorities, and hence constitute 'implicit' rather than 'explicit' policies.

This chapter reports the results of two separate but related studies which bear on these issues. The core of the paper is the analysis, presented in the third section (which is pitched at the sectoral level) and the fourth section (which is based on data on individual firms) of information collected in the course of an intensive field study carried out in Egypt in late 1981. The main focus of attention is the extent to which Egyptian Government policy conditions technology choice and development. The analytical framework within which the central discussion is set is, in part, the outcome of the other component of the study, which comprises an investigation of the extent to which factors related to market structure and industrial organisation (MSIO) influence the choice of technology. The latter component forms the basis for the second section, in which is presented a summary of previously published evidence on the MSIO-technology relationship in eight LDCs together with an account of the results of a new examination of the position in a further eight countries.

Unusually among developing countries, Egypt, for long a classic 'dual' economy with a large reservoir of unemployed labour, currently enjoys a low level of unemployment. Available estimates put the 1979 figure at 4–5 per cent and are described by the recent ILO Employment Mission as 'relatively satisfactory'.[1] However, while it is certainly possible that the current unemployment situation will persist, or even improve, in the foreseeable future, the presence of certain significant factors in the labour market does provide grounds for pessimism. Specifically, the probability of a decline in the level of activity in the construction industry, the increasing participation rate of women in the labour force, the possibility of repatriation of a significant proportion of the 0.7 to 1.5 million Egyptian nationals working in other Middle Eastern countries should the boom there fade, and the increased reluctance of the public sector to act as an 'employer of last resort' – all give cause for concern.

This being so, it seems reasonable to argue that the employment effects of technology policy are by no means irrelevant, but may, indeed, assume considerable importance in years to come. This is not to say that any investigation of technology policy in Egypt should deal exclusively with the issue of labour intensity versus capital intensity, but such a discussion is clearly of central importance. Accordingly, an underlying assumption in the analysis of the third and fourth sections, and in the concluding passages (fifth section), of the present study is that, at some stage in the not too distant future, the Egyptian authorities may wish to use technology policy to generate additional jobs, both by fostering (selectively) 'inherently' labour-intensive industries (a policy recommended by the ILO Mission as central to 'hedging against possible set-backs') and by encouraging the substitution of labour for capital where the nature of production functions permits. Efficiency in such an undertaking demands a clear understanding of the various ways in which, both deliberately and, so to speak, incidentally, current policies affect employment levels via technology choice, together with an awareness of the scope for manipulating technology appropriately through such policies. Subsequent sections of this study are addressed to these issues.

TECHNOLOGY CHOICE: MULTI-COUNTRY STUDY

As a preliminary to the main analysis, it is helpful to examine as broad as possible a spectrum of potential influences on technology choice. The factors determining the level of capital intensity in the firms in a given economy – and here we focus on manufacturing enterprises – are not, of course, restricted to those directly or readily amenable to central government influence. Many other forces come into play, and it is important that the more significant of these are identified and understood if we wish to assess with any accuracy the independent impact of policy variables.

Broadly speaking, the principal factors likely to influence choice of technology, aside from any specific government directives on the matter, fall into three categories: production function constraints, factor market characteristics, and the complex of considerations earlier termed MSIO.

The role of the first two of these is well established. With regard to production functions, other things being equal, it is to be expected that observed levels of capital intensity will be associated with the magnitude of the elasticity of substitution (low values of the elasticity going with high capital intensity), with the range of capital intensities available, and with the strength of scale economies (powerful economies of scale usually being realised through the application of capital-intensive methods). As to factor market conditions, it is generally accepted that variation in relative factor prices across industries or sectors is likely to set up pressures encouraging variation in technology (higher wage rates (w) and/or lower rental costs of capital (r) tending to favour the substitution of capital for labour).

The case for expecting to find some relationship between MSIO and the factor proportions embodied in manufacturing technology is a theoretical innovation of fairly recent vintage. The central argument, attributable to White (1976), is that while, in a perfectly competitive world, the firms in low-wage economies will be driven by the 'policing' action of market forces to choose labour-intensive technology, the existence of market imperfections makes this less certain. Assuming that a preference for high technology *per se* appears in the utility function of technology choosers in the 'managerial firm' (in either private or public sectors), it is to be expected that deviations from perfect competition and the use of 'inappropriately' capital-intensive technology will go together.

In practice, the influence of market structure cannot be measured directly, but may be represented by the separate effects of the *degree of concentration* (industries with greater degrees of concentration being more highly monopolised and thus particularly prone to excessive capital intensity, other things being equal), and the *strength of foreign competition* (the greater the import penetration of a given market, the weaker is monopoly power).

The MSIO factors determining the level of capital intensity in the firms in a given economy are not, of course, restricted to those relating to the degree of competition or monopoly, but extend beyond this to embrace two further potentially important structural characteristics – the extent of state involvement in production and the presence of multinational corporations; both tend to be associated with a predilection for high technology.

All these influences on factor proportions may be expected to affect the capital intensity of manufacturing technology across industries in the direction indicated in Table 6.1, and are incorporated in the analysis.

It will be appreciated that in fact several of these factors may themselves be interrelated in greater or lesser degree. In particular, it is likely that the extent of state participation and the scale factor will be positively correlated, as public enterprises are often established because of market failures – which tend to be more frequently encountered in industries characterised by large-scale, capital-intensive production.[2] Again, greater multinational corporation participation may be expected to go along with the existence of scale economies. And since the potential for large-scale operation tends to be positively related to the degree of concentration (the incompatibility of falling long-run average cost curves and competition is well known), the possibility of further, complex relationships amongst the 'influences on capital intensity' listed in Table 6.1 opens up.

For purposes of the present study, a review was made of available evidence on the MSIO-technology interaction (in all, ten studies, covering eight countries were located), and this was supplemented by analyses of census of production material for a further eight LDCs.

For each of the sixteen countries covered, the econometric procedure adopted in attempting to 'explain' the choice of manufacturing technology was to regress some, or all, of the variables listed in Table 6.1 on the dependent variable – a measure of capital intensity. In all cases ordinary least squares (OLS) techniques were applied to cross-sectional data drawn mainly from national industrial census reports.

TABLE 6.1 *Hypothesised influences on capital intensity of technology*

MSIO characteristic	Measure	Direction of association with capital per man
Level of concentration	Output of 3 largest establishments as % of total output of industry	+
Degree of import penetration	Imports (cif) as % of domestic output	–
Extent of state participation	Output of state establishments as % of output of industry	+
Extent of multinational corporation (MNC) participation	Output of MNCs as % of output of industry	+
Elasticity of substitution	Index of technical rigidity*	–
Strength of scale economies	Value added per establishment	+
Factor–price ratio	Average wages over average interest rates	+
Maximum capital intensity	Fixed assets or value added per employee in the corresponding industry in the United States	+

NOTE* For the definition of this concept, see Forsyth, McBain and Solomon (1980).

For most countries the overall explanatory power of the independent variables taken together was considerable; the performance of individual regressors, in terms of the sign attached to, and the statistical significance of, their coefficients, is summarised in Table 6.2. Panel 1 of the table refers to the earlier studies, and panel 2 to the new analyses.

Perhaps the best known of the studies referred to in panel 1 is the analysis of the case of Indonesia (in Wells, 1972) in which the principal finding was that increasing degrees of monopoly tended to be accompanied by subordination of the profit-maximising goals of 'economic man' to the high technology preferences of 'engineering man'. Here the objective function of owners of firms enjoying monopolistic advantages was said to include 'Reducing operational

TABLE 6.2 *Summary of findings on MSIO-technology choice*

[Dependent variable: Capital intensity of manufacturing technology] Country	Independent variables								
	Extent of state participation	Level of concentration	Scale	Factor price ratio	Elasticity of substitution	Degree of import penetration	Extent of foreign participation	Maximum capital-intensity	R^2
(1) Published findings									
Indonesia[1]	+						+		
Brazil[2]		+	+	+			+		
Pakistan[3]	+	N/S	+	+					
Ghana[4]					N/S	N/S	N/S		
Ghana[5]						N/S*	+#		
Ghana[6]					–				
Turkey[7]			+		–	–*			
Malaysia[8]			+ +		–				
Philippines[9]			+ +		–	N/S*	–		
Kenya[10]									
(2) Survey results[11]									
Chile		–	+	+	+	N/S		N/S	0.90
Sri Lanka	N/S	N/S	N/S	N/S	N/S				0.00
Hong Kong		+	+						0.29
Singapore		N/S	N/S						0.08
Korea (Rep. of)		N/S	+	–	N/S	N/S		N/S	0.21
Cameroon		N/S	N/S		N/S	N/S		N/S	0.35
Thailand		N/S	N/S		N/S	N/S		N/S	0.00
SAC		N/S	+	N/S	N/S+		+	+	0.90

NOTES
(i) + Significant at 5% level or better (F-test); interpretation of positive and negative relationship as in Table 6.1.
(ii) N/S Non-significant
(iii) * In these cases the higher levels of protection and lower levels of import penetration are assumed to go together.
(iv) # The picture at the level of the individual industry was more complex than this overall view suggests; in some industries foreign ownership was associated with lower capital-intensity.
(v) SAC= a South-East Asian Country.

SOURCES 1.,3 Wells (1972); 2. Newfarmer and Marsh (1981); 4. Baah-Nuakoh (1980); 5., 6., 7., 8. Forsyth, McBain and Solomon (1980); 9. Pack (1976); 10. Forsyth and Solomon (1977); 11. Survey carried out as part of present study.

problems to those of managing machines rather than people. . .
Producing the highest quality product possible. . . [and] Using sophisticated machinery that is attractive to the engineer's "aesthetic"'.
Wells also found that foreign ownership, state participation, increased wage rates and/or reduced interest rates, and possession of
foreign patents tended to go along with the use of capital-intensive
technology. Although his sample was too small and fragmented
(forty-three firms drawn from six different industries) to permit
rigorous analysis of what is clearly a complex multivariate relationship, Wells' work provides useful impressionistic support for the
various hypotheses advanced in the previous subsection.

The studies by Newfarmer and Marsh (1981) and White (1976)
employ a more rigorous approach to the analysis of the issues
addressed by Wells. Both attempt to 'explain' observed choice of
technology across a range of industries in a single developing country
in terms of a number of the variables adduced above. Newfarmer and
Marsh, examining the choice of technology in Brazilian industry,
found that approximately 42 per cent of the variation in observed
capital–labour ratios was attributable to variation in the level of seller
concentration (a 4-firm ratio was used), nationality of ownership,
scale and value added-to-sales ratio. They concluded that 'the more
oligopolistically structured a market and hence muted the forces of
price competition, the less compelled are firms to utilise the lowest
cost combinations of factors'. For Pakistan, White conducted a
similar analysis, again finding the level of seller concentration to be
closely associated with the level of capital intensity, and concluding
that ' a competitive environment, whether created from internal or
external sources, does appear to encourage (technical) flexibility in
socially worthwhile directions in Pakistan and likely in other LDCs as
well, i.e., in labour-intensive directions'.

However, in the case of Ghana, Baah-Nuakoh (1980), using data
at the level of the individual firm, was unable to establish significant
relationships between capital intensity and concentration, import
penetration, foreign participation or elasticity of substitution but did
find that larger scale, higher wage/rental costs of capital ratios, and
enhanced state participation all tended to go hand in hand with more
capital-intensive production.

No other studies focusing directly on the overall MSIO-technology
nexus have been located, but a limited amount of additional evidence
is available for some countries with respect to certain aspects of the
relationship. Specifically, Forsyth, Solomon and McBain,[3] in the

course of an examination of the influence of 'technical rigidity' on factor proportions in Turkey, Malaysia, the Philippines and Ghana, found the monopoly-favouring characteristics of substantial levels of tariffs on imports to have a clear-cut positive effect on domestic capital intensity only in the case of Turkey. However, *a priori* expectations were more clearly borne out by the conclusion that higher levels of capital intensity were associated with higher levels of scale, and with lower values of the elasticity of substitution.

Finally, analyses of the influence of nationality of ownership on technology choices have yielded very mixed results. For example, studies of this point in Kenya[4] and Ghana[5] suggested that, in the former country, foreign owned firms were relatively labour intensive, while in the latter the opposite was true, although in this case marked variation was observed across industries. The conflict of evidence here is fairly typical of the several empirical investigations of this particular point, though both Helleiner[6] and Lall,[7] on reviewing the contradictory evidence, come to the conclusion that, on balance, multinational corporations tend to use relatively capital-intensive techniques.

Turning now to the new evidence summarised in panel 2 of Table 6.2, it is necessary first to note that this is based on variable definitions which in some cases diverge from those used in the earlier studies (the data were drawn from national census of production reports), though these divergences might be expected to affect coefficient values rather than their signs (the only exception to this being the 'elasticity of substitution' for which a redefinition of the variable leads us to anticipate a positive rather than a negative sign on the coefficient).

The results are rather 'disappointing' in that they provide little support for the majority of the hypotheses advanced earlier. The overall 'explanatory' power of the regressors varied considerably from country to country, being high in the cases of Chile and SAC, and very low in the cases of Sri Lanka and Thailand. Very few statistically significant relationships were encountered among individual regressors, the obvious exception being the scale variable, for which five of the eight countries displayed, as before, a positive relationship. Furthermore, an examination of the country-by-country results indicated a consistent tendency for scale to be the dominant regressor.

Although detailed data were available on the 'level of concentration' in all eight countries, only in the cases of Chile and Hong

Kong were the regression coefficients significantly different from zero (at the 5 per cent level on an F-test), and the direction of association was, for Chile, contrary to *a priori* expectation. Surprisingly, variation in the elasticity of substitution was not found to be closely related to variation in capital intensity in the majority of countries in the sample, though the anticipated (positive) relationship was identified in the cases of Chile and SAC (in one of its two data variants). The outcome of the analysis of the influence of variation in factor prices was largely inconclusive, and no significant results were obtained for the impact of variation in the degree of import penetration. Regrettably, satisfactory data were available on neither the 'extent of foreign participation' nor the 'extent of state participation', and these potentially important influences on the capital intensity of manufacturing technology were thus largely excluded from the analysis.

Taken together with the findings of the earlier studies referred to in Table 6.2 the results indicate that, while very little of the available evidence conflicts with the expectations represented in Table 6.1, only in the cases of scale and, to a lesser extent, elasticity of substitution, is there strong and fairly consistent support for the hypotheses advanced. It does seem to be the case that the attraction of cost reductions accruing from the application of capital-intensive technology at higher levels of scale, combined with the bounds set on factor substitution by elasticity of substitution configurations, are the dominant influences on choice of factor proportions. Compared with the impact of these production function-determined characteristics the independent influence of market signals, competitive pressures and market power – embodied in the other variables – seems weak or non-existent.

Given the problems inherent in interpreting cross-country results based, inevitably, on highly aggregated data drawn from heterogeneous sources, and the important possibility of the existence of multicollinearity – between, in particular, the scale, state participation, foreign participation and concentration variables – it is perhaps inadvisable to put a great deal of weight on this 'ultra-determinist' finding. However, the general configuration of the results does provide us with a useful lead-in to our analysis of technology choice in Egypt, indicating, at the very least, the need to pay particular attention to the role of certain variables, and providing us with a broad, multi-country context within which to set the results of the Egyptian study.

TECHNOLOGY POLICIES AND PREFERENCES IN EGYPT

In this section a review is made of the various 'explicit' and 'implicit' elements of Egyptian Government technology policy, and an interpretation offered of the reaction of private sector technology choosers to the partial insulation from market pressures which they enjoy as a result of government policies.

The discussion of policy falls into three parts: the first deals with broad, 'explicit' policies and the general direction of their effects; the second is devoted to an examination of the probable effects on technology choice of a series of 'implicit' policies; and the third is concerned with the effects of one specific aspect of policy – the setting up of Free Zones. While the discussion of the Free Zones should be seen as self-contained, the material presented in the first two subsections is intended as a prelude to, and a means of generating a number of hypotheses to be examined in, the analysis of data and information assembled in the course of the survey of Egyptian manufacturers reported in the fourth section. (It is not, however, feasible to use direct surveys of manufacturers' 'motivation' as a means of assessing accurately the impact of *all* identified elements of technology policy, either because of data limitations, or because policy measures are not always perceived as impinging on individual firms – having a diffuse, 'contextual' effect rather than an immediate impact on producers. So our analysis of the consequences for technology choice of the various policies adumbrated is partly a matter of empirical investigation, partly *a priori* speculation.)

Finally, the 'technology policy' of the private sector is examined in the fourth subsection.

'Explicit' Technology Policies and Their Consequences

Over the last two decades, the Egyptian Government has accepted a considerable part of the responsibility for guiding and developing the economy. Until the early 1970s, the 'commanding heights' of the production sector were firmly occupied by the state and almost all industrial investment was in the public sector. A sizeable proportion of all technology choice decisions was thus taken within the public sector, and the view of that sector on what constituted an 'appropriate' choice was of paramount importance. Since the early 1970s, the introduction of the 'Open Door' ('Infitah') policy, which encourages the growth of the private sector (including the foreign-owned sector) as part of a general move away from *dirigisme*, has weakened the direct influence of the state on the process of technology selection while increasing the importance of other participants in this process. Nevertheless, the public sector accounts for around 75 per cent of investment, and the policies and attitudes of that sector may be expected to play a crucial role in conditioning the behaviour of the private sector.

Few LDC governments have promulgated clearly articulated technology policies, and still fewer have constructed the framework of legislation and administrative infrastructure required to give such policies meaning. Egypt is no exception in this respect, and explicit statements of government policy on technology are few in number. Such comments as have been made are, however, clear in their support for the installation of high technology.

Thus the crucial Law 43, which is publicly held to be the corner stone of the Open-Door Policy, has among its main aims 'Creating favourable conditions for technology transfer' and 'Emphasising the priority of projects that have the ability to increase foreign exchange earnings and bring most up-to-date technology'.

This emphasis on modern, sophisticated (and, inevitably, capital-intensive) technologies reflects an entrenched, if largely uncritical, faith in the potential of the latest equipment. (Official antipathy towards the importing of second-hand machinery is a corollary of this.) Since the early days of the modern public sector it has been embodied in the planning and appraisal methodology used by state enterprises. Thus Hansen notes that 'Egyptian Ministries, packed with engineers, have often been accused of being biased in favour of capital-intensive projects, this being confirmed by their eclectic approach to project appraisal in which . . . the total investments needed for each sector were estimated by the application of sectoral capital-output coefficients, obtained from historical studies of domestic development, from engineers, and from experience abroad.'[8]

High technology predilections have also, it seems, affected policy-makers' attitudes to the *timing* of the establishing, or expanding, of industries. As in most LDCs, in Egypt a foundation of traditional handicraft and small-firm activity in textiles, leather, and wood products existed long before the emergence of the state-dominated industrial sector, 'but it is characteristic of modern Egyptian industry that it did not grow organically from this basis'. What has happened, rather, is that domestic demand was allowed to build up to quite large dimensions – satisfied by imports – until the market was big enough to permit the establishment of a fairly large-scale industry, and 'This is true for all the pioneering industries, sugar, cement, cotton ginning and pressing, spinning and weaving, and chemical fertilisers.' [9]

A consequence of this approach to industrialisation is that the processes of investing in plant and machinery and of creating jobs have not been seen by the government as inextricably intertwined, but rather as separate issues. Obviously, other things being equal, more investment has meant more jobs, but the possibility of generating employment on a large scale by manipulating technology does not seem to have been considered, or, at any rate, given a high priority. By and large the unemployment problem has, until recently, been ameliorated by the expedient of encouraging overmanning in both industry and public administration. The giving of guarantees of employment to graduates and former soldiers is now being phased out, but was for long a prize example of a system of job creation whose *modus operandi* was the creation of barely concealed, suppressed 'redundancy' on a massive scale.

All in all, it is perhaps not surprising that, in a society in which science is held in high regard, and in which scientific education is much esteemed (two-thirds of all graduate degrees are in pure and applied sciences, and there are over 260 research institutes – many attached to Ministries) – 'engineering man' should hold sway in the public sector, particularly at a time of relatively rapid expansion of output and employment. As yet, the potential dangers of the situation have not aroused much internal debate, but the most recent thoroughgoing analysis of Egyptian science policy stated the problem in unambiguous terms: 'Many of Egypt's current political and economic policies are predicated on the belief that a major infusion of Western technology and know-how will produce rapid growth, modernisation and widespread benefits to the population . . . However . . . investment based heavily on imported technology . . . could skew income sharply and visibly and favour capital-intensive modes of manufacturing to the detriment of employment.'[10]

'Implicit' Policies

Almost any economic or social policy is likely to have *some* implication, however indirect, for technology choice, and a problem of selection for analysis arises. For present purposes, five such policies have been identified which are associated with distortions in internal and external sector price signals, and a further five reflecting disparate aspects of official intervention. All are, at least potentially, significant influences on technology choice.

(a) *Wage policy*: Real wage rates more than doubled over the period 1970–9. Much of the increase was attributable to the 'quickening' of economic activity over those years, but the government also played an active, and significant, role in the labour market, increasing minimum statutory wage rates regularly, requiring the payment of sizeable increases in bonuses, pensions and other labour benefits, and permitting public sector employers to pursue a 'wage-leader' policy.

It is to be expected that the upward pressure on real wages (ahead of productivity growth) will encourage substitution of capital for labour where other factors do not supervene; experience in other LDCs suggests that the distorting power of this factor will be greatest in the medium-to-large scale formal sector enterprises, and weakest in small-scale and informal sectors.[11]

(b) *Cost of capital*: There is no *general* shortage of investment funds in Egypt. Remittances from Egyptians working abroad, foreign aid, inward foreign investment, oil and Canal revenues provide substantial blocs of capital for industrial borrowers – a circumstance tending to favour capital-for-labour substitution. Government policy here has been largely passive, though there has been some lending to the private sector at below market rates.

(c) *Price controls*: Maximum price controls on consumer goods may be expected to discourage production of goods subject to controls while encouraging production of other goods.[12] An analysis of the production functions of eighty-one commodities subject to price controls suggested that the majority had relatively higher potential than the goods to which demand is likely to have been shifted.

(d) *Foreign exchange*: The Egyptian Government has, in recent years, encouraged, through exchange controls and export promotion, the appreciation of the £E. Since most capital equipment is imported, this has markedly cheapened investment in plant and machinery.

(e) *Tariffs*: The existence of a protective tariff on imports may be expected to influence technology choice in Egypt in two distinct ways. Within-industry choice may be affected, in the manner discussed in the second section by the removal of the 'policing' effect of market forces. At the same time, differential tariff protection may promote resource reallocation from industries with lower levels of protection to the more heavily protected industries – a process which

TABLE 6.3 *Effective rates of protection of Egyptian manufacturing*

Industry	ERP	Industry	ERP
Beverages	7.94	Non-electrical machinery	0.30
Electrical machinery	5.86	Oil and coal products	0.29
Transport equipment	3.54	Chemicals	0.24
Wood and products	3.39	Paper	0.01
Garments	0.44	Spinning and weaving	−0.53
Rubber and products	0.43	Non-metallic products	−0.80
Basic metal	0.38	Food	−0.92
Tobacco	0.37	Leather and products	−3.69

SOURCE   M.K. Eldin and R.E.B. Lucas, *Comparative Advantage in Egyptian Manufacturing* (Cairo: USAID, 1981).

may increase or reduce overall capital intensity, depending on the factor proportions prevailing in the industries affected.

The second of these effects was examined by relating the level of effective protection (ERP) in seventeen major Egyptian industrial sectors to a measure of capital intensity given by the ratio of labour costs to non-labour (see Table 6.3 for ERP values).

The outcome of this analysis was the conclusion that, within the industries examined, the pattern of protection is biased very slightly in favour of labour-intensive industries, though this becomes a rather stronger net bias in the opposite direction when the industries are weighted by current levels of labour use. However, it is not possible to put much weight on this result as the observed relationships were not statistically significant.

(f) *Standards legislation* regarding both product quality and safety provisions is likely to encourage mechanisation.

(g) The desire of the Investment Authority (GAFI) in earlier years to *avoid 'excessive' overlapping*, or duplication, of productive capacity (in order to minimise excess capacity) has almost certainly had the effect of increasing seller concentration; this policy has now been relaxed, though, according to some interviewees, not completely abandoned.

(h) *Employment policies* which forced the absorption of surplus labour by public sector enterprises seem certain to have exerted a depressing influence on capital intensity in those manufacturing concerns affected.

(i) The strong encouragement given to *exporting* usually implies improved product quality, and thus improved technology, if foreign competition is to be met.

(j) *Strategic considerations*: Madkour notes that 'The weapons industry was no doubt one of the major areas . . . for the transfer of modern sophisticated technology to Egypt',[13] and it is clearly the case that mass production of modern armaments in Egypt is likely to involve standardised high-precision work only feasible with capital-intensive processes. This effect spills over into the non-military sector since much of the military productive capacity has been expressly designed to be convertible to civilian production.

Free Zones

As an important part of the 'Open Door' policy, the Egyptian Government began, in 1972, accepting applications from investors for permission to invest in industrial sites designated as 'Free Zones'. Firms in these zones operate under preferential conditions in the form of tax holidays, relief from restrictions on repatriation of profits, relief from customs duty on imports of capital goods and material inputs, and guarantees against nationalisation or confiscation of assets.

Both public sector and private sector Free Zones have been set up, the initial aim being to attract foreign investment though most of the privileges accorded to Free Zone firms are now available to indigenous firms under the terms of Law 43. By 1977 no fewer than 740 projects had been approved by the Investment Authority – 206 in Public Free Zones, 62 in Private Free Zones, and 472 in the 'Inland' sector, although the number of projects actually operating as late as 1980 was still much smaller than this.

In view of the reduction in risk and the removal of foreign exchange restrictions, it was to be expected – and it was intended – that the bulk of projects coming forward would be from foreign-owned firms (or joint ventures involving such firms) or from outward-looking, foreign market oriented Egyptian firms. Indigenous firms not wishing to purchase foreign capital equipment would be expected to display less interest in the potential advantages than would the former categories of firm.

Thus the specific terms of the privileges (which provide an implicit or explicit subsidy to investors and importers of capital goods), and the anticipated character of the firms likely to wish to take advantage of them, both imply that the introduction of the new regulations will bias capital intensity upwards in projects conducted under their conditions.

The figures in Table 6.4 indicate that, on average, Free Zone establishments employ nearly eleven times as much capital per man as the larger non-Zone establishments. In none of the five industrial sectors for which a comparison can be made is the Free Zone capital intensity less than twice that of the non-Zone establishments. This is, in part, the result of the particularly capital-intensive character of the foreign-owned enterprises which have been attracted to the Free Zones.

These figures indicate very clearly that the setting up of the Free Zones by the Egyptian Government has fostered the growth of a new, particularly capital-intensive sector in the economy. (It should be noted, however, that only parts of this new sector represents investment 'diverted' from the rest of the economy. The special conditions applicable in the Zones will almost certainly have 'created' new investment for Egypt by attracting foreign-owned firms which would otherwise have gone elsewhere, and, to a much lesser extent, by stimulating indigenous investment.)

Competitive Pressure and Technology Preferences

It can be argued that, over a broad span of Egypt's industry, the special circumstances under which technology is chosen are unlikely to encourage the clearly defined, systematic variation envisaged, in the model constructed in the second section, of the capital intensity of technology with the extent of state control and the level of concentration respectively.

In a purely statistical sense, this is partly explained by the lack of variability in the level of state control in those industries for which information is available. In twelve of the eighteen largest manufacturing sectors, state enterprises are responsible for over three-quarters of all output, and in seven of the eighteen sectors they account for over 90 per cent. Moreover, in all of those industries in which state participation is substantial, it seems likely that there will be interaction between the public and private sectors – that is to say, their choices of technology will not be made independently of one another. Private sector firms in these industries are inevitably left

TABLE 6.4 *Capital intensity in industries in the Free Zone and in other private sector industries, 1977*

Industry	(i) Free Zone	(ii) Other large- scale private	(iii) Other small- scale private*
Textiles	39.43	3.21	3.21
Food and beverages	23.40	8.30	5.25
Chemicals	182.96	7.07	5.34
Engineering	15.91	3.39	6.65
Metallurgical	31.11	3.92	2.96
Mining	9.00	3.98	2.50
Electrical	NA	0.92	1.63
Leather	NA	5.71	3.45
Average	50.29	4.56	3.87

NOTE NA – Not Available *All having more than 9 employees.
SOURCE (i) Government Authority for Investment and Free Zones, *Report on Arab and Foreign Investment*, Cairo, 1978.
(ii) and (iii) World Bank: Arab Republic of Egypt: *Survey of small scale industry*, Washington DC, 1977, and Government Authority for Investment and Free Zones (GAFI).

with very little room to develop a competitive milieu. Much more common is the situation in which private quasi-monopolies operate side by side with state quasi-monopolies, both largely immune from the pressure of the market on their technology-choice processes, the commercial weakness of much of the state sector reinforcing the insulation of the private sector.

What of those industries in which the state sector is less dominant? In the majority of these sectors, most of which are less 'concentrated', it can still be argued that 'competitive' pressures on technology choice are likely to be very weak – because of the special circumstances under which the 'modern' private sector has sprung into life in the recent past. The Egyptian private sector has, in the last few years, operated under a set of special, non-competitive circumstances (beyond those already discussed in relation to MSIO) which may be expected to have affected profoundly its attitude towards technology choice, in most cases encouraging the use of high technology. Thus:

(a) The emphasis (partly government inspired) on rapid growth of output may be expected to have favoured the use of capital-intensive methods, which yield quick returns in terms of their impact on growth rates of production.

(b) The erosion of the 'policing' property of market forces is almost certainly exaggerated by the fact that many new private firms deal largely with the public sector (a reflection of the dominant position of the latter) which is itself not subject to strong competitive pressures.

(c) It has been suggested that a significant proportion of new private investment in the early days of the liberalisation policy had as its source 'Those who had made their private wealth from public enterprises (and) were. . . . anxious to find safe outlets for it.'[14] Certainly, the change in policy released pent-up investment, some of it no doubt financed in this way, some from funds becoming available after the legislation of 1974 put an end to 'sequestrations' and 'enabled individuals to reacquire some of their previously confiscated private properties (which) . . . added wealth to the local private sector and set it completely free.'[15] It seems, moreover, that much of the new investment was carried out by individuals with a background in trading, who had little prior knowledge of production technology. All of these factors would appear to favour investment in 'safe', established capital-intensive methods of production, rather than what might be perceived as a lengthy, and perhaps futile, search for a more 'appropriate', more profitable labour-intensive technology.

(d) The small size of local markets for consumer goods, the bias of local consumer goods markets towards higher quality variants of products (this reflecting Egypt's highly skewed income distribution), and the sharp decline in the early 1970s of sales to Eastern European countries – have all favoured high quality production for local or sophisticated Western markets. Since, in many industries, high quality output and high technology go together, this effect may be expected to have biased technology choice towards more capital-intensive methods.

All these pressures point away from the need for firms to feel constrained to concentrate on 'fine-tuning' of technology choice to local factor availability, and have, clearly, had a considerable effect on thinking about technology, to the point where the 'conventional wisdom' in the medium-to-large firms in the private sector clearly favours a high technology policy. The existence of a private sector 'technology policy' would, of course, be entirely redundant in a world of static competition. In such a world the dictates of profit-maximising motivation would guide each firm to the most appropriate technology (in 'private' terms). But in a rapidly expanding private sector fostered by a government which is acutely conscious of its role as standard bearer of new and revolutionary (in local terms) economic policies, and which is subjected to unambiguous propaganda regarding the 'proper' direction technology policy should take – it is not surprising that the kind of plant and machinery employed has become an objective in itself, independently of its implications for profitability.

The view that it is desirable to use sophisticated technology as an end in itself – or, at least, as an end not clearly related to economic advantage – was voiced frequently in the course of interviews conducted with firms in the modern sector. And it is expressed frequently in the reports by individual industries in the main publication of organised Egyptian industry – the *Year Book* of the Federation of Egyptian Industries[16] – thus: '*Plastic* plants have tended to modernise their products by using the latest in modern technology' (p. 14); the *rubber* industry has made progress with 'modernisation of its machinery in keeping with modern technology' (p. 24); the 'most important . . . projects include modernisation of the *soap* industry' (p. 49); 'the *cotton ginning* industry is keen to raise the standard of that industry through a programme to develop the existing mills and establish new ones based on the most up-to-date means' (p. 3); 'the *dyeing, printing and finishing* industry has of late witnessed fundamental change which upgraded production and raised the quality to world technical standards . . . due to (use of) especially modern machines' (p. 38); in *bread* production the Federation endorses the view that 'facilities of all sorts' should be made available 'to capitalists who would like to establish full or semi-mechanised bakeries' (p. 12); and so on for virtually every industry.

This strong preference for high technology is emphasised by the views of the Egyptian scientific community which on the whole favours the emergence of a modern, private, capital-intensive manufacturing sector. This bias goes hand in hand with an almost total absence of interest in R&D effort aimed at identifying or extending the labour-intensive sector or production functions, and a general dependence on imported machinery and ideas, and was widespread in the larger firms covered in our survey. The recent Lyman Report endorses this interpretation: 'Industries have . . . come to depend heavily on imported technology, and the access to technological expertise that comes with a foreign licence. There appears to be little incentive . . . for making a special effort to develop or adapt Egyptian technology in place of imported . . . Dependence on outsiders for problem solving (is a) cultural characteristic of Egyptians.'[17]

All of this applies specifically to the medium- to large-sized private firms and to state enterprises. The smaller firms in the private sector tend to have very much less contact, in terms both of commercial transactions and personal interaction between managements, with the state sector than do their larger counterparts. Moreover, many of the factors bearing on the technology choice of the latter group – wage norms, price controls, foreign exchange controls, quality standards, and so on – are likely to have less relevance to the small firms. In short, competitive pressures can be expected to be much stronger in the small-scale sector, so it is probable that, *ceteris paribus*, firms in that sector will tend to use relatively labour-intensive technology.[18]

The previous subsections have been devoted to an analysis of those aspects of Egyptian Government policy which seem likely to influence the choice of technology in both state and private sectors, and to a review of attitudes towards technology in state and private manufacturing and in the scientific and technological research sector. All the evidence adduced indicates the existence of widespread biases towards the use of capital-intensive technologies.

We now go on to examine how the more accessible of these policy-determined biases operate at the motivational level – that is, how they are perceived by those responsible for decision-making with regard to technology, and how far they affect decisions. This approach, while lacking the breadth of coverage in industry-level analysis employed in the second section, and the facility provided by the multivariate approach used in the second section for (under certain circumstances) isolating the respective effects of the 'independent' factors bearing on technology choice, does in this case have the distinct advantage of going directly to the subjective level at which decisions are taken, hence permitting an unravelling of various influences otherwise difficult to separate because of their multicollinearity characteristics.

TECHNOLOGY POLICY AND TECHNOLOGY CHOICE IN EGYPT – THE MICRO-ECONOMIC EVIDENCE

In the first subsection below the results of a series of interviews carried out with Egyptian manufacturers are reported. Given the dominance of the scale factor noted in the second section (and confirmed below), the central question to be answered with the aid of the survey data may be formulated as follows: Does the clear difference observed in choice of technology in small-scale and large-scale

sectors simply reflect production functions characterised by a tendency for the productivity of capital relative to that of labour to rise as
scale increases (and thus for profit-maximising capital intensity at
given factor prices to rise with scale, other things being equal)? And
does this effect swamp pressures which might be expected to lead to a
greater variety of factor proportions across firms and industries (e.g.,
variation in wage rates and cost of capital) especially in industries
with 'flexible' technologies? Or is the relatively high capital intensity
of the medium/large firms a (possibly inappropriate) choice sustained
by the effects of certain of the implicit or explicit government technology policies outlined in the third section?

More specifically, the focus of attention in this section can be
summarised as follows: (a) The *nature* of the choice of technology
made in the firms interviewed; (b) The *role of government policy* in
promoting, or permitting, the observed choices; and (c) The *appropriateness* of these choices.

Technology Choice in the Firm

The Sample Firms

The population from which the sample of firms to be interviewed was
drawn comprised published listings of manufacturers (supplemented
by information provided by various institutional sources)[19] in four
broad industry groupings – engineering, leather products (including
footwear), metal products, and spinning and weaving. The sample of
firms was drawn by a stratified–random procedure. In all, forty-two
firms provided sufficiently detailed information to warrant inclusion
in the study. Where possible, the actual individuals involved in the
process of technology choice were interviewed, generally the managers or owners in the smaller and medium-sized firms, and engineers
and managers in the larger concerns. All interviews were carried out
in late 1981.

The principal characteristics of the firms in the sample are indicated in Tables 6.5 and 6.6. It will be seen that twenty-three of the
firms had twenty or fewer employees, that is, were 'small' firms
(which somewhat under-represents such firms in terms of units while
over-representing them in terms of numbers of employees). With
reference to capital intensity, the majority of cases (twenty-nine, or
71 per cent) had fixed assets with an estimated replacement cost of
less than £E5000 per employee at 1981 prices, and ten firms were

TABLE 6.5 Main characteristics of the sample firms

Industry	Number of employees				% Employees unskilled/semi-skilled				Capital intensity*			
	1-10	11-20	21-150	over 150	0-25	26-50	51-75	76-100	below 1000	1000-4999	5000-9999	over 9999
Leather products	4	4	2	2	4	5	1	2	4	6	2	–
Electrical eng.	–	2	2	1	2	–	2	1	–	3	1	1
Mechanical eng.	–	3	–	1	2	2	–	–	2	1	–	1
Metal products	2	3	4	2	5	1	2	3	2	5	4	0
Spinning and weaving	3	2	2	2	3	1	5	–	2	4	2	1
Total	9	14	10	8	16	9	10	6	10	19	9	3

* £E/employee

TABLE 6.6 Capital intensity by industry and scale in firms surveyed

| Industry | Average capital intensity | | Skill mix | |
| | Small firms | Medium/ large scale firms | Small firms | Medium/ large scale firms |
	(£E/employment)		(Labour force skilled and semi-skilled)	
Leather products	1 022	4 311	29.4	64.1
Electrical & mechanical engineering	2 350	7 196	21.6	50.2
Metal products	2 855	5 120	56.2	24.8
Spinning & weaving	3 001	6 183	40.8	47.3
Total	2 161	5 753	35.9	48.3

clearly at the labour-intensive end of the factor-proportions spectrum, with values of less than £E1000 per employee. As regards skill intensity, it seems clear that, overall, higher values tend to be associated with the smaller scale of operation – a characteristic reflecting, in part at least, indivisibilities at the managerial and supervisory levels, and the tendency of the small-scale sector, in certain cases, to merge into the 'craft' sector.[20] The observed skill profiles are of clear relevance to the 'appropriateness' issue, and are returned to below.

Choice of Technology

Characterisation of technologies was carried out in two different ways for purposes of the present study, respondents being asked first to provide data on the replacement cost of fixed assets (plant and machinery only), and then to allocate their technology to one of four classes – advanced/standard practice (in industrialised countries)/obsolete/very simple. The first approach suffers from a number of well-known estimation problems, the most intractable of which are non-availability of certain older-vintage technologies (in Egypt one set of specialist sewing machines still in occasional use in a footwear factory was said to date back to before the First World War) and lack of precision on the part of interviewees in specifying current prices of presently available capital goods. Thus the figures derived from the survey are probably best regarded as representing orders of magnitude rather than precise estimates.[21]

The values given in Table 6.6 indicate that the range of (average) factor proportions observed was wide, running from a high of £E7196 per employee in engineering to a low of £E1022 per employee in leather products. The highest, and lowest, values recorded for individual firms were, respectively, £E12 516 for a mechanical engineering firm and £E109 for a shoemaker. It will be noted that these values fall well below the Free Zone averages reported in Table 6.4. In fact, only one of the firms in our sample was currently operating under Free Zone regulations.

Policy and Choice of Technology in the Firm

In this subsection, the information assembled on 'motivational' influences is examined seriatim, and then simultaneously in a multivariate analysis.

Scale and Factor Proportions

All available data suggest that the small-scale private sector of Egyptian manufacturing is much more labour intensive than both the remainder of the private sector and public enterprises. The figures quoted in one official source,[22] indeed, indicate that in every one of the twenty main sectors of Egyptian manufacturing the smaller firms were relatively labour intensive, the overall labour-to-capital coefficient for these firms being twice that for the larger firms.[23] This is, of course, a familiar pattern. What is of particular interest in the present context is the extent to which it is attributable to policy elements as opposed to straightforward production function characteristics.

As the figures quoted in Table 6.6 indicate, in our sample average capital intensity in each of the industries covered is markedly higher in the larger firms. Further, a simple bivariate test did in fact indicate the existence of a significant positive association, and this result is confirmed by the outcome of classification by firms of their technologies in terms of the four-way 'sophistication' index. Of the larger firms, three (16.7 per cent) classed their technologies as 'advanced', nine (50.0 per cent) used 'standard practice' technology, six (33.3 per cent) used 'obsolete' technology, and none admitted to using 'very simple' technology. The corresponding per cent figures for the smaller firms were 'advanced' – 4.3 per cent, 'standard practice' – 34.8 per cent, 'obsolete' – 39.1 per cent and 'very simple' – 21.7 per cent. Thus the positive relationship between capital intensity and scale encountered earlier is reproduced in our sample of firms.

Cost of Labour

It has already been argued that Egyptian government policy (implicit and explicit) on factor prices is likely to impart an upward bias to capital intensity in manufacturing. Wage data from our sample of firms indicate clearly that the smaller firms tended to pay substantially lower manual labour wage rates than did the larger firms – the overall margin (skill-adjusted) being 37 per cent. It was also clear that, as in many other LDCs, the larger firms believed that statutory wage provisions bore much more heavily on them than on the less 'salient' smaller firms; the latter group did not disagree with this view. As to the impact of labour costs on choice of technology, the rising unit cost of labour (wages and associated benefits) encouraged by legislation and by the active wage-leadership of the public sector, was frequently quoted in the medium/large-scale sector as an important pressure encouraging the substitution of capital for labour; increasing capital intensity was seen as a means of 'escaping' the serious upward pressure on costs resulting from rising labour remuneration. Most respondents in the larger firms spontaneously cited labour costs as having a bearing on technology choice, and almost all (17 out of 19) confirmed the existence of a positive influence when a roster of 'potential influences' was introduced for discussion. Among smaller firms this effect was reported as being very much weaker, though most interviewees did invoke rising labour costs as a reason for *wishing* to mechanise further.

In terms of an apparent correlation at the sectoral level, then, it seems that growing unit labour costs, to some extent stemming from government intervention in the labour market, were positively related to capital intensity. Moreover, firms right across the scale spectrum perceived the relevance of labour costs to the choice of technology, though only the larger firms claimed to have been influenced in practice. This having been said, it must be added that it is not possible on the basis of qualitative data alone to ascertain the extent to which perceived pressures – here of wage rises on technology – actually affect decisions.

Cost of Capital and Capital Goods

As regards the *cost of capital*, the *de facto* subsidies on imported equipment available to larger firms through preferential access to already overvalued foreign exchange (estimated to be equivalent to a 10 per cent reduction in capital costs at the time of our survey), the remission of import duties on capital goods (on the recommendation of GAFI – worth approximately 12 per cent on capital costs in 1981), and the availability of tax holidays (which reduce the cost of investing by releasing a relatively cheap flow of internally generated funds) – all clearly favour the purchasing of more capital-intensive technology. In the public sector, this is compounded by the long history of virtual isolation, now breaking down, from cost of capital considerations, such that, freed from market constraints, technology choosers have been able to select relatively 'risk-free' and aesthetically satisfying high technologies in precisely the way anticipated by the 'managerial discretion' hypothesis referred to in the first part of the second section.

Amongst the firms interviewed in the course of the study, experience varied considerably as to the ease with which investment finance could be raised, and the extent to which restrictions on the availability of such finance acted as a deterrent to increasing capital intensity. Once again, the largest firms displayed, on the whole, characteristics sharply distinct from those observed in the smaller-scale sector. Of the eight firms with over 150 employees, none reported capital shortage or the cost of capital as a factor significantly affecting their investment decisions. Of the nine 'medium-sized', three reported problems over, and delays in, obtaining the foreign exchange required for importing capital equipment, and had, as a result, either improvised or delayed purchase of the equipment. (This appeared to be a problem of long standing.)

The small firms in the sample displayed a virtual unanimity of experience regarding the availability of finance and its effect on their technology. Contrary to the views of a number of commentators on this issue, these firms referred to problems in gaining access to institutional finance and to short-term credit, and, in particular, to the unwillingness of commercial banks to make advances against the kind of collateral available to such firms. Despite the efforts of the development bank (DIB), it seems that small firms in Egypt face a situation familiar in LDCs, in which they are forced to rely on internally generated finance or high-cost borrowing from money-lenders.

A further disadvantage affecting the small firms in our sample was the fact that the 40 per cent or so of their plant and machinery which was imported did not benefit from the cost reductions available through GAFI; foreign exchange for such equipment had ultimately to be purchased on the 'free market'.

In general, the impact of cost of capital considerations was perceived as much more severe in the small-scale sector than in the larger-scale sector, to the point where firms in the former were found to be operating a substantial proportion of fairly aged equipment, kept going by the exercising of considerable ingenuity in terms of repair and local manufacture of components, and even of complete machines on an *ad hoc* basis.

Market Power

As our earlier attempts to examine the influence of market power using 'concentration' measures yielded rather mixed results, a somewhat more sophisticated approach was adopted in the survey. Specifically, information was gathered on (a) the geographical coverage of the market (proportion of output of main product sold locally, nationwide, and in the export market); (b) firms' perceptions of the strength of overall competition faced by the main product (in terms of, where possible, an estimate of market share – or, otherwise, placing on a three point intensity scale: weak/intermediate/strong–and information on the strength of quality, as opposed to price, competition); in addition, firms were asked directly what effect, if any, competition had on their approach to selecting plant and machinery.

A majority of firms were unable to give any convincing quantitative estimate of market shares held, so this measure had to be dropped. The thirty-six satisfactory responses on the alternative measure of the degree of competitiveness were collapsed into a two-way measure, and cross-tabulated against the principal market for output identified by each firm.

For each of the six-way classifications, the number of firms, the average values of capital intensity, and the number of employees were computed.

The outcome of the analysis of these data was negative in that it proved difficult to establish the existence of a clear link, for the whole sample, between the degree of monopoly power, as perceived by the individual firms, and the capital intensity of their technology. The average capital-per-man for firms facing 'strong' competitive pressure was, in fact, slightly higher (though this was a non-significant result) than the value for firms facing weaker competition (and hence having greater market power) – contrary to our expectation.

Most of the eight firms which exported, including three of the four reporting facing weak/moderate competition, emphasised the need to achieve high *quality*, and, as a secondary consideration, some noted the value of a capacity for quick response in exploiting export markets (this with reference to flexibility of output levels); these considerations were seen as requiring the use of sophisticated technology. The operation of this 'quality' factor could not, however, be demonstrated for production for the broad 'national' market rather than the immediate 'local' market. The importance of the quality–technology relationship was, in fact, noted only in relation to the export market (and to import competition discussed below).

Finally, it was found that the exporting firms, as well as being more capital intensive than the non-exporters, were significantly larger, averaging 260 employees as against thirty-seven, this again emphasising the collinearity of size and other factors associated with high capital intensity.

Tariffs and Import Competition

Tariff levels and competing imports appear to have had a very considerable effect on technology choice, but not in the direction anticipated. Instead of increased competition of this sort forcing firms back onto the profit-maximising, labour-intensive sectors of their production functions, the opposite has happened, and in the industries examined an entirely different mechanism seems to be at work. A majority of the firms in the engineering and metal products industries[24] reported that tariff levels low enough to permit imports, i.e., low enough to encourage competition, led to a need for the use of more sophisticated technology to permit improvement in product quality.[25] Indeed, GAFI was reported as having intervened in a small number of cases to advise applicants for concessions that their proposed technologies were insufficiently sophisticated to meet import competition, and should be reconsidered. Firms having less interest in the 'sophisticated' end of the market (these being, for the most part, smaller than the others) by and large did not experience a quality-induced bias to their technology choice.[26]

The overall figure for capital-per-employee used in the import-competing firms was more than double that in the other firms, and was marginally higher than the average value for medium-sized and large firms recorded in Table 6.6.

Price Controls

Price controls on consumer goods sold on the domestic market were not, in general, felt by firms to have much bearing on the choice of technology,[27] though they were reported by some firms as having the effect of pushing them into exporting, with the consequences for technology choice outlined above.

Explicit Government Policy on Technology

The Egyptian Government has for many years favoured increased sophistication of technology and has communicated this preference to the medium/large-scale sector. Although job creation is numbered among the criteria applied by GAFI in processing applications for concession, it does not appear to be a particularly heavily weighted factor. And although GAFI favours the use of 'second-ranking'

TABLE 6.7 *Distribution of sources of technological know-how by capital intensity and size of firm*
(% of firms)

Source of technological 'know-how'	Capital intensity		Size	
	low*	high	small	medium/ large
1. Foreign firms granting licences or providing technical assistance	12.5	41.2	13.6	36.8
2. Suppliers of capital equipment (foreign and local)	16.7	23.5	9.1	31.6
3. Own technical staff/ own *R&D*	12.5	11.5	13.6	10.5
4. Owner's know-how	33.3	11.8	40.9	5.3
5. Customers	4.2	5.9	0.0	10.5
6. Traditional know-how	12.5	0.0	13.6	0.0
7. Government technical assistance; scientific and technological institutions	8.3	5.9	9.1	5.3
Total	100.0	100.0**	100.0**	100.0

NOTES   * Having plant and machinery per employee of less than the average for the sample (i.e. less than £E3738).
        ** Error due to rounding.
SOURCE   our survey.

technology rather than the very latest equipment, nevertheless countervailing considerations such as the pressure for high quality production[28] tend to dilute this approach in practice. The state sector itself has a long history of heavily capital-intensive investment, often on a turn-key basis because of the lack of indigenous R&D and innovatory expertise, and has tended to pull many private firms in the same direction because of the close input–output relationship which exists between the two sectors. None of these pressures was reported as having much relevance to the small-scale sector; indeed, the EIDDC frequently advises the firms with which it deals to favour labour-intensive technology.

The Availability of Skills

This was a matter of widespread concern. The outflow of skilled labour to other Middle Eastern countries in the recent past has placed serious strains on the labour market, and shortages of specific skills (artisans, technicians, supervisors) were reported by virtually all of the firms in our sample. Nevertheless, few of the firms were willing to invest significant sums in training because of the fear that trained men would soon be lost to foreign employers. The frequently reported response by large firms to this dilemma was recourse to machinery in place of men. The shortage of investment finance in the small-scale sector, however, usually prevented adoption of such a strategy there, though shortage of skills was felt keenly by small firms.

The Technology Choice Process

The process by which technology is chosen by the firm may itself constitute an important influence on the nature of that choice, independent of the various influences, or potential influences discussed above. The criteria applied in judging suitability, the thoroughness of the search for a suitable technology, and the specific sources of information available and drawn on, are all potentially significant.

In practice, it proved difficult to separate out the criteria peculiar to the technology choice process from many of the factors mentioned above. This was, clearly, because technology choice fulfills a variety of functions, both overt and underlying, certain of which the firm itself may not fully appreciate. A full analysis of the fundamental motivation of individual firms – that is, of the nature of the owner's and/or management's objective functions – and of the nature of the linkage of such motivation to technology choice, involves very detailed study of individual cases, and cannot be accomplished in the course of a one-off interview. In the present study the approach adopted was to focus primarily on the factors 'external' to the firm, i.e., those factors reported on earlier, but to probe the nature of the decision-making process in terms of that aspect of it which seemed most amenable to objective analysis – the sources of information drawn on by the firm.

Respondents were asked to rank seven possible sources of technological know-how in order of importance to their particular firm with regard to the currently installed plant and machinery. Information on the *principal* source of information in each case is presented in Table 6.7, with a breakdown by capital intensity and size. It will be seen that, not unexpectedly, the orientation of relatively capital-intensive firms is towards foreign (mainly developed country) suppliers of know-how (as such), and towards machinery suppliers (most of which were also located abroad, in developed countries); two-thirds of the relatively capital-intensive firms derived the information on which they based their choice of technology from such sources, which might be expected to be heavily biased towards recommending/supplying high technology. Know-how generated within the firm, from own R&D or information possessed by technical staff and/or the owner, was the main source of information in only just under one quarter of the relatively capital-intensive firms. The labour-intensive firms, in contrast, were much more heavily oriented towards internal sources of expertise and 'traditional know-how' (i.e., freely available 'conventional wisdom' on simple techniques) with nearly two-thirds of all firms favouring these sources. Relatively few firms, even in the labour-intensive sector, gave the government's technical advisory schemes or Egyptian scientific and technological institutions as the principal source of technical information. Nor were customers of primary importance in this context to either group of firms.

Turning to the subdivision of the sample by scale, a similar pattern of information-sourcing emerges, this reflecting the fact that, with the exception of a small number of fairly sophisticated firms in the small-firm bracket, size and capital intensity are clearly aligned in our sample.

Overall, then, the picture is one in which larger, more capital-intensive firms display much greater 'outward' orientation in deriving their technological know-how than do smaller and more labour-intensive firms. It seems inevitable that the source of their information and the level of sophistication of their technologies will be closely correlated, and the figures presented in Table 6.7 seem to bear this out.

Why has this pattern of orientation of a key element in the technology-choice mechanism emerged? It seems that both 'internal' and 'external' factors are at work. By and large, firms with an outward-looking policy in know-how gathering referred to the pressure of various factors mentioned earlier – the need for improved product quality, the desire to export, shortages of skilled labour, high wages, and so on. At the same time, virtually all of the large enterprises visited had no effective means of generating their own, 'appropriate', technology, and looked beyond Egypt for sources of information on production and management methods, often in the form of joint ventures with foreign firms. Local scientific effort was held to be 'too theoretical', and as the machine-producing industry is still in its infancy, the bulk of 'know-how' and equipment had to be imported. This was very clearly the case in the entrepreneur-dominated firms, in which management itself was short of technical skills, but even in firms managed by engineers, external sourcing of technology was the norm.

Despite the useful extension efforts being made by the government with regard to small-scale technology, most small firms interviewed had little knowledge of any recent developments in labour-using technologies which might be available to them and suitable for their operations. A few reacted to the need to refurbish or renew equipment by carrying out their own modifications or minor innovations, others purchased second-hand machines from larger firms, or new machines from abroad. Overall, this imparted a mild, but continuous upward pressure on capital intensity.

Moreover, in general no objective preference for labour-intensive technologies was expressed by such firms; on the contrary, small manufacturers using primitive or obsolete equipment by and large wished to re-equip with modern machinery as and when circumstances permitted.

A Discriminant Analysis of the Policy-Technology Choice Relationship

Discriminant Analysis and Hypotheses

Our earlier analysis suggests that firms choosing capital-intensive and labour-intensive technologies, respectively, are likely to operate under quite different conditions. The examination of individual policy-influenced factors above has indicated that those factors which appear to be associated with technology choice are scale, wage rates, access to capital, competitive pressure from imports, and sources of technological information. It is of interest to consider the *simultaneous* influence on choice of technology of such of these as can readily be quantified, given our data base.

Since the firms interviewed, rather than lying scattered along a spectrum with respect to capital intensity and to the various influences on technology choice, tended to be bunched at either end of the range, it is likely that the association of particular factors with the choice of technology is more appropriately examined using discriminant analysis than the regression techniques we have thus far employed.

Using discriminant analysis, it is possible to assess the efficiency with which a set of 'discriminators' (here 'influences' on technology) actually 'separate' one group of firms from another (here capital intensive from labour intensive).[29] The discriminators used were capital intensity (measured as £E/employee), scale (number of employees), wage rate (£E/employee p.a.), market power (entered as a binary variable taking the value 1 in cases in which competitive pressures were reported as being 'weak' or 'moderate', and 0 for 'strong' competition), and competition from imports (a second binary variable to be interpreted in the same way as that representing market power). The average level of capital intensity in the corresponding industry in the United States was also included in the analysis, as before. The hypothesis to be tested is that high capital intensity in Egypt goes along with large-scale production, relatively high wage rates, stiff competition from imports and high capital intensity in the United States, but that the market power/technology relationship is weak.

Results

The results of the discriminant analysis are presented in summary form in Table 6.8. They indicate that, as predicted, scale, wage rates and the level of import competition are all useful in separating firms with high and low capital intensity, respectively. In all cases the relationships are positive. The first three discriminators are significant at the 5 per cent level (or better)[30] but the last named is non-significant. The sign on the coefficient of the 'market power' variable is negative, but this result is also non-significant, as is the case for US capital intensity. These results may be interpreted as indicating that capital-intensive firms in our sample may be distinguished from relatively labour-intensive members of the sample in terms of three characteristics – scale, wage rates paid, and extent of import competition. As the classification matrix (panel 3, Table 6.8) shows, these three measures alone can correctly classify 75 per cent of the sample firms. Thus we may infer that the factors associated with the choice of technology in firms which opt for capital-intensive technology are systematically different from those at work in other firms, and different in ways consonant with the findings of the earlier discussion.

TABLE 6.8 *Summary of results of discriminant analysis: capital intensity and MSOI characteristics*

(1) *Basic statistics* Groups	*Discriminators*				
	Scale	*Wage-rates*	*Monopoly power*	*Imports*	*US capital intensity*
(i) High capital intensity					
mean	137.7	586.11	0.44	0.50	11 673
standard deviation	176.5	256.94	0.51	0.51	2 981
(ii) Low capital intensity					
mean	38.43	436.09	0.46	0.09	9 996
standard deviation	50.76	170.45	0.51	0.29	3 184

(2) *Discriminant analysis*
   (a) Canonical correlation: 0.47   Overall significance level: 5% (F-test)
   (b) Canonical discriminant function coefficients:

Discriminator	*Coefficient*	*Significance level (F-test)*
Scale	0.64676	5%
Wage-rates	0.05695	5%
Market power	0.34927	non-sig. at 5%
Imports	0.89691	1%
US capital intensity	0.06996	non-sig. at 5%

(3) *Classification matrix*

Actual group	*No. of cases*	*Predicted group membership*	
		High cap. int.	*Low cap. int.*
High capital intensity	17	11 (64.7%)	6 (35.3%)
Low capital intensity	19	3 (15.8%)	16 (84.2%)

Per cent grouped cases correctly classified = 75.0%

Capital intensity and 'appropriateness'

Discriminators and hypotheses

Having considered the relationship between MSIO factors and technology choice in Egypt, it is now desirable that some means be provided of assessing the level of 'appropriateness' of the choices of technology observed in the course of the sample survey of Egyptian firms. We have observed that the presence of a number of MSIO factors, some of which might properly be regarded as 'distorting' competitive market relationships, has been associated with a tendency for firms to choose relatively capital-intensive technologies – that is, technologies more capital intensive than might otherwise have been chosen.

This is of interest in itself, but, though it implies that such choices may be inappropriate, it is not logically necessary that this is so. What is required here is a *separate* test of appropriateness. A rigorous analysis would, of course, require very extensive investigation of both economic and engineering aspects of the wide range of technical alternatives available in the industries in question. This is far beyond the scope of the present study. However, if we are to move beyond simply observing that high capital intensity is often associated with inappropriate choice, it is necessary to attempt to develop some other criteria of 'appropriateness'. It is emphasised that this attempt is of a speculative nature, but it *is* found to tell us something of interest about the observed pattern of technology choice of the firms in our sample and, by extension, about the impact of the MSIO factors considered in the earlier analysis.

Once again, bearing in mind the discontinuities mentioned earlier in the relationship between policy-influenced variables and technology choice – such that the groups of more, and less, capital-intensive firms, respectively, constitute separate sectors – it is apparent that discriminant analysis is a suitable analytical tool for examining the proposition that high capital intensity and relative 'inappropriateness' go together.[31]

The assessment of the degree of appropriateness, or otherwise, was carried out by reference to the incidence of high, and low, capital intensity groups of firms, of key factors generally held to be associated with this characteristic in terms of private or social rates of return. Specifically, it was argued that 'excessively' high capital intensity would result in (i) *operating problems* due to the excessive sophistication of plant and machinery given local management, labour and raw materials. This problem would be expected to manifest itself in a variety of ways, the most amenable to measurement being 'down-time' due to operating problems. 'Down-time', our first discriminator, was measured by the proportion (percentage) of total nominal capacity (working days) lost through breakdown, including time lost due to non-availability of spares; (ii) a significantly *greater demand for scarce skills*, skilled labour being a resource in particularly short supply in Egypt. Skill intensity, our second discriminator, was measured by the ratio (percentage) of unskilled and semi-skilled direct and indirect workers to total employees; and (iii) *a tendency to generate relatively weak backward linkages* to indigenous industry – due to the requirement for sophisticated inputs not produced locally – which is undesirable from both growth-transmission and balance of payments points of view. The strength of backward linkages, our third discriminator, was defined as the proportion (percentage by value) of all material inputs purchased locally, and believed to be of local origin.

A heavier incidence of down-time, a greater demand for scarce skills, and a more pronounced tendency to generate relatively weak backward linkages are all characteristic consequences of inappropriate choice of technology, and were hypothesised to distinguish relatively capital-intensive from relatively labour-intensive firms.

Results

The principal results of the discriminant analysis are presented in Table 6.9.

It will be seen (panel 1) that the relative mean values for all three of the 'appropriateness' measures (discriminators) in the two groups of firms are in the predicted configurations. The relatively labour-intensive firms in group (2) display very tight grouping round the mean values for 'down-time' (with a standard deviation of only 5.51 from a mean value of 10.23) and for 'linkages' (with a standard deviation of only 28.65 from a mean of 88.93). The same is true, though to a lesser extent, of group (1) firms.

The figures in panel 2 indicate that although the *overall* discriminatory property of the three variables taken together is high (significant at the 5 per cent level on an F-test), only the 'down-time' and 'linkages' variables are individually significant (again at the 5 per cent level); the 'skill-ratio' does not emerge as an effective discriminator.

These results indicate that firms can be classified into high, or low, capital intensity groups using values for 'down-time' and 'linkages' as classificatory criteria (higher values for the former and lower for the latter being associated with higher capital intensity, as predicted). The observed variability of the skill ratio is, however, such as to rule it out as an efficient discriminator.

TABLE 6.9 *Summary of results of discriminant analysis of 'appropriateness'*

(1) *Basic statistics*

groups	Discriminators		
	Down-time	Unskilled/ semi-skilled %	Linkage
Group 1 – high capital intensity			
mean	18.92	44.80	69.20
standard dev.	22.91	58.66	37.89
Group 2 – low capital intensity			
mean	10.23	30.13	88.93
standard dev.	5.51	25.91	28.65

(2) *Discriminant analysis*

   (a) Canonical correlation 0.511   Overall significance level 5%
                                    (F test)
   (b) Canonical discriminant function coefficients:

discriminator	coefficient	significance level (F test; d.f. = 3,37)
down-time	0.1912	5%
skills	0.3043-01	non-sig at 5%
linkage	−0.2949-02	1%

(3) *Classification matrix*

Actual group	No. of cases	Predicted group membership	
		(1)	(2)
(i) High capital intensity	21	7(33.3%)	14(66.7%)
(ii) Low capital intensity	20	15(75.0%)	5(25.0%)

Per cent grouped cases correctly classified = 70.73%

Overall, as is shown in panel 3 of Table 6.9, the discriminators can classify some 70 per cent of all cases accurately, considerably better than the *a priori* expectation of a 50 per cent success rate.

The results may be interpreted as indicating that the choice of high technology has associated with it a systematic tendency for the incidence of operating problems (here measured by down-time) to be relatively high, and for the strength of linkages to the domestic economy to be relatively low. These results can be construed as providing support for the view that technology embodying high capital intensity tends to be less 'appropriate' than less sophisticated technology; given that both explicit and implicit technology policies in Egypt have been found to favour enhanced capital intensity, this subsequent finding has obvious implications for the framing of future technology policy.[32]

INTERPRETATION AND CONCLUSIONS

The aim of this study has been the elucidation of the relationship between government policies and technology choice, special attention being paid to characteristics of market structure and industrial organisation which are affected by policy and which in turn condition technology.

A review of the limited literature bearing on the MSIO/technology relationship indicated that the dominant influence on choice in the countries studied was *scale* of production. A preliminary review of the situation in Egypt suggested that, here too, scale of production was the key determinant of factor proportions. However, while this finding seemed, at first sight, to relegate other MSIO factors (and government influence) to relative insignificance, the possibility remained that non-scale factors were, in fact, of importance, but were scale-specific, that is, that they did affect choice of technology, but in different ways at different levels of scale. Clearly, complex patterns of this sort are difficult to identify using standard regression techniques applied to sector-wide or even industry-wide data. Application of such techniques, it seems, almost always yields an explanation of technology choice which depends on the largely production function – scale interaction.

Our study of the Egyptian case, the methodology of which permits access to a much richer vein of information on the nature of the investment process than is attainable using published data at the industry level, suggests an alternative, more subtle explanation which may well apply, to a greater or lesser extent, in other LDCs. What seems to be happening in Egypt is that most medium- to large-sized firms in both the private and public sectors tend to aim for the highest feasible degree of sophistication in their new technology, irrespective of the level of concentration, wage rates, the elasticity of substitution, and import competition. Apparently, opportunities for increasing profits by taking advantage of factors favouring more labour-intensive investment (e.g., the availability of low-cost[33] labour and 'flexibility' of the production function at the target level of output) are simply ignored, regarded as irrelevant. In this drive towards higher capital intensity, production function constraints setting maximum levels on attainable technology constitute the boundary which firms in the sector cannot cross, though this boundary extends as scale levels rise permitting further augmentation of capital intensity.

But this is not to say that policy (and MSIO) factors are unimportant. Indeed, they are crucial to the survival of the process, and to the creation of the dual structure of technology observed in Egypt. It is by a particular combination of policy-determined MSIO characteristics that the competitive pressures and the consciousness of scarcity are observed to hold capital intensity down in the small-scale sector; equally, a somewhat different combination of MSIO characteristics fosters higher capital intensity in the larger-scale sector.

The behaviour of the latter sector may be characterised as one of collective collusion to opt for high technology, with the government making this possible for most larger firms (even in otherwise competitive industries) by submerging the pressures which would militate against this and replacing them by an environment favourable to the use of high technology. Tax and tariff subsidies to the purchases of high technology, preferential access to cheap finance and to cheap foreign exchange (together with an overvalued currency), penalties imposed on the greater use of labour, encouragement of higher production for local and export markets, deliberate attraction of foreign investment, the build-up of a conventional wisdom in government (but embracing industry and the scientific community) favouring the use of advanced technology, and a general arranging of economic policies to permit profitable survival of firms taking the high technology route – all these features of Egyptian economic policy combine to create a 'plateau' upon which it is possible for large numbers of firms to operate at 'artificially' high capital intensities. Fierce competition at this level is quite consistent with a stable equilibrium choice of factor proportions far removed from that implied by the structure of true opportunity costs. Indeed, paradoxically, strong competition with foreign (and even indigenous) suppliers in domestic and export markets is a powerful force *promoting* the use of high technology in the many industries in which product quality and sophistication of techniques are positively related. [34] Removal of the protective MSIO structure would immediately put pressures, similar to those now experienced by the smaller firms, on many of their larger counterparts. [35] In addition to such within-industry effects, there would also be a shift of resources back towards the more labour-intensive industries (part of which would take the form of a withdrawal, in due course, from at least some of the industries in which the quality–technology link is strong).

The experience of the smaller firms in Egypt, which operate in a much less 'distorted' form of competitive environment than do larger firms, tends to support this view. And further confirmation, though of a fairly tenuous kind, may be seen in the finding that positive identification of a significant relationship between the level of concentration and technology choice was virtually restricted to larger LDC economies.

Thus our conclusion is not that pressures on the choice of technology flowing from the character of the central government's policies and the MSIO structure they foster have been irrelevant or inoperative in Egypt, but rather that policy has favoured the emergence of two distinct sets of MSIO characteristics which have pushed factor proportions apart in the smaller-scale and larger-scale sectors of industry, respectively, so creating a growing polarisation – a 'dual' industrial sector. This has been due largely to explicit, and implicit, Egyptian Government policy on a number of key MSIO variables, to the point where a degree of 'inappropriateness' may have been engendered in the large-scale sector. A reversal of these policies might be expected to have important consequences for labour absorption in many industries – an important consideration if the predicted slackening of the labour market comes to pass, for it seems that not only choice of industry, but also *within*-industry choice of technology has been, and can be, affected by policy. (This finding, which gives grounds for optimism regarding the leverage, in terms of employment creation, of appropriately modified policies, differs from the conclusion of the ILO Mission, which saw choice of industry alone as the key to the creation of new employment opportunities).

However, the evolution of a set of technology policies designed to promote job creation will certainly not happen spontaneously. Until now a very large and influential segment of the totality of Egyptian Government policy has gone unrecognised as such by the authorities. The prerequisite for an effective reorientation of existing measures is an understanding at all levels of policy-making that 'implicit' technology policies exist – and may have profound effects. It must be appreciated that, *inter alia*, exchange rates, wage levels, tax regimes, tariff rates, the cost and availability of investment finance and capital equipment, and the strength of competitive pressures in both private and state sectors of manufacturing – all of which may be influenced by government decision – are all, at least potentially, important in determining, via their impact on technology choice, the rate of job creation in Egyptian industry.

It is true that the capacity already exists, at least in a formal sense, to monitor and manipulate technology choice, through GAFI, in firms applying for 'concessions'. It may thus be suggested that this provides an additional degree of freedom to government in pursuing policies which might otherwise have undesirable side-effects on the technology profile. This is, however, a dangerous line of argument to pursue. There is no evidence that official intervention in this area has done much more in the past than delay investment, and GAFI possesses neither the financial resources nor the very broad-based expertise required to provide an adequate technology advisory service, let alone a system of control. Under these circumstances it is clearly preferable to permit enterprises a free choice of technology and to seek to condition such choice by manipulating the 'environment' in which choosers operate in ways suggested by our earlier analysis.

How easy such a reversal of policy would be in Egypt, or in other LDCs where similar pressures apply, is another matter. Socially inefficient technology policies – implicit or explicit – which give a heavy weighting to the pressures towards high capital intensity stemming from scale economies and product quality considerations, are widespread and resistant to change. They persist largely because of a combination of vested interest, ill thought out (or non-existent) technology policy, and the existence of significant elements of 'X-inefficiency' and 'high technology preference' on the part of technology choosers. Correcting all of these is likely to be a major undertaking, and a continuing rather than a once-for-all process. The evidence from our sample survey was that the preference for high technology is very widespread and that fairly firm constraints are required to keep 'engineering man' in check. (This applies not only to the manufacturing sector, but also to the scientific and technological 'establishment', which clearly prefers to be associated with 'off-shore' high technology than to concentrate on technology development and modification at a level capable of being assimilated by the Egyptian capital goods sector.)

While LDC firms may not from preference seek to operate in an environment which reflects the full realities of economic life in a low-income country, it is desirable that they should do so. In the context of the Egyptian economy, this means that the various elements of market structure and industrial organisation which at present so strongly favour the use of capital-intensive technology should continue to be sustained by the government only if it is clear that there are long-term dynamic benefits accruing to such a policy, and these outweigh the costs of apparent misallocation of resources, foregone opportunities for job creation, and the institutionalisation of the dual industrial society.

NOTES

1. Hansen and Radwan (1982). A detailed description and analysis of the working of the Egyptian labour market is given in chapters 3 and 4.
2. For evidence on this point relating to the Republic of Korea see Jones and Mason (1979).
3. Forsyth, McBain and Solomon (1980).
4. Pack (1976).
5. Forsyth and Solomon (1977).
6. Helleiner (1975).
7. Lall (1978).
8. Hansen (1965).
9. *Ibid*.
10. Lyman *et al.* (1980). With regard to planning and technology, the ILO Mission (Hansen and Radwan, 1982) notes that the current 'Plan does in fact assume capital-intensive investment across the board' (p. 180).
11. The ILO Mission found that the statutory minimum wage rates 'apply in principle to all public and private employment, but are, outside the public sector, effective mainly in larger enterprises' (p. 147) – and also noted that 'The Government's wage policies have in the past been almost exclusively equity-oriented' (p. 15) rather than concerned with allocative efficiency.
12. In so far as price controls are offset by government subsidies this effect will be diminished. Subsidies are important in the cases of food products and textiles, accounting for, respectively, 12 per cent and 10 per cent of the gross value of sales.
13. Madkour (1980), p. 218.
14. Mokhtar (1980).
15. *Ibid*, p. 70.
16. Federation of Egyptian Industries (1980).
17. Lyman *et al.* (1980).
18. I.e., in terms of the original market structure hypothesis; production function effects of scale are discussed later.
19. Particularly important among these institutional sources of information were GAFI (the Investment Authority – which handles requests from firms for tariff and tax concessions, access to foreign exchange on

preferential terms, etc.), the Engineering and Industrial Design Development Centre (responsible for small-scale industries' extension services), the Industrial Development Bank, the Central Agency for Public Mobilisation and Statistics (CAPMS), the Federation of Egyptian Industries (which provides technical and administrative assistance to its 5000 members – these including all public sector enterprises), the Chamber of Engineering Industries, the Chamber of the Leather Industries, the Chamber of Spinning and Weaving Industries, the Chamber of Food Industries, the General Federation of Egyptian Chambers of Commerce, the American University and USAID.

20. The very smallest establishments visited in the leather products and metal products industries were closer in kind to the traditional handicraft producers than to 'modern' manufacturers.

21. It is very often the case, in smaller establishments, that fixed assets falling under the heading 'land and buildings' are difficult to value since premises may not have been purpose built, may comprise all, or part, of the owner's residence, may be used for different purposes at different times, and so on. Hence the value of plant and machinery is preferred as a measure.

22. CAPMS (1970).

23. At 4.930 employees per £E'000 value added as against 2.179.

24. The number of firms reporting strong import competition in their product lines in leather products and spinning and weaving was insufficient to provide a sufficiently broad base for comparison.

25. The highly skewed distribution of income in Egypt, with attendant well-developed tastes for high quality goods amongst the upper income brackets, reinforces this effect.

26. Though some small firms did report a somewhat different quality-related phenomenon in the shape of inadequate or defective machinery adversely affecting product quality.

27. The impact on the allocation of resources between industries is, of course, a separate matter.

28. One Ministry of Industry official observed that setting detailed standards' specifications could be tricky, but that the problem was easily solved in cases in which the product was already produced in industrialised countries, 'Since we can then apply the foreign standards.'

29. For a detailed discussion of the logic of discriminant analysis see Forsyth and Solomon (1977).

30. On an F-test.

31. Note that it is *not* assumed here that a high capital–labour ratio *per se* constitutes inappropriateness, though this is a common assumption elsewhere. Given current conditions in Egypt's capital market (at least as they affect the larger firms) it is not clear that the social cost of capital is much above the market price. Under the circumstances, a superior test of appropriateness of existing technology choice is an operating one, which refers to the extent of the assimilation of the technology, low levels of assimilation indicating inappropriateness. (Tests of future technology choices would, of course, take employment into consideration if conditions in the labour market deteriorated.)

32. It must be emphasised, however, that this analysis of appropriateness should not be regarded as an adequate study. The data base is small, and the approach adopted essentially static. A very much more detailed, product-by-product study would be required before anything definitive could be said.
33. I.e. relative to that in developed countries.
34. Though this is not true of all industries. For an example of the converse involving the Egyptian carpet industry see El-Karanshawy (1975).
35. In many cases the effect may be expected to correspond to a shift from 'quality competition' to price competition.

REFERENCES

Baah-Nuakoh, A. (1980) *Factor Use and Structural Disequilibrium in a Developing Economy* (Reading: University of Reading), a Ph.D thesis.
CAPMS *Quarterly Survey of Industrial Production* (Cairo: various years).
Forsyth, David J.C. and Solomon, R.F. (1977) 'Choice of Technology and Nationality of Ownership in Manufacturing in a Developing Country', *Oxford Economic Papers*, July.
Forsyth, David J.C., McBain, N.S. and Solomon, R.F. (1980) 'Technical Rigidity and Appropriate Technology in Less Developed Countries', *World Development*, May–June.
Hansen, B. (1965) *Development and Economic Policy in the United Arab Republic* (Amsterdam: North-Holland).
Hansen, B. and Radwan, S. (1982) *Employment Opportunities and Equity in Egypt* (Geneva: ILO).
Helleiner, G.K. (1975) 'The Role of Multinational Corporations in the Less Developed Countries' Trade in Technology', *World Development*, September.
Jones, L. and Mason, E. (1979) *The Role of Economic Factors in Determining the Size and Structure of the Public Enterprise Sector in Mixed Economy LDCs*, Discussion Paper No. 61 (Boston: Department of Economics, Boston University).
El-Karanshawy, H. (1975) *Choice of Carpet Weaving Technology in Egypt* (Glasgow: Strathclyde University), a Ph.D thesis.
Lall, S. (1978) 'Transnationals, Domestic Enterprises, and Industrial Structure in Host LDCs: A Survey' *Oxford Economic Papers*, July.
Lyman, P. *et al.* (1980) *US Co-operation with Egypt in Science and Technology* (Cairo: USAID and Academy for Scientific Research and Technology).
Madkour, N. (1980) *Egyptian Defence Programmes and their Impact on Social and Economic Development of Egypt* (Cairo: American University), an MSc thesis.
Mokhtar, N. (1980) *The Upper Economic Class in Egypt* (Cairo: American University), an MSc thesis.
Newfarmer, R.S. and Marsh, L.C. (1981) 'Foreign Ownership, Market Structure and Industrial Performance', *Journal of Development Economics*, February.

Pack, H. (1976) 'The Substitution of Labour for Capital in Kenyan Manufacturing', *Economic Journal*, March.

Wells, L.T. (1972) *Economic Man and Engineering Man: Choice of Technology in a Low Wage Country*, Economic Development Report No. 226 (Cambridge, Mass.: Development Research Centre).

White, L.J. (1976) 'Appropriate Technology, X-Inefficiency, and a Competitive Environment: Some Evidence from Pakistan', *Quarterly Journal of Economics*, November.

7 External Development Finance and the Choice of Technology

JOHN WHITE[1]

INTRODUCTION

In the period in which aid[2] began to emerge as an important element in relations between developed and developing countries, in the 1950s, it was seen by donors, and more especially by the United States in its relations with its strategic allies in Asia as an instrument of foreign policy.[3] In the early 1960s, perceptions changed. With the initiation of the First Development Decade, the establishment of the OECD's Development Assistance Committee (DAC) in 1961 and the holding of the first United Nations Conference on Trade and Development (UNCTAD) in 1964, the Western donor countries' bilateral aid programmes and the multilateral agencies' programmes came to be seen as part of a worldwide and collective development effort. But there was still not a very precise view of what it was, in developmental terms, that aid was intended to achieve. Donors tended simply to assume that development was identical to growth of national product, and that they were providing resources which in one way or another would contribute to accelerated growth. A more refined collective donors' view was presented explicitly for the first time in 1968, in the DAC report for that year. It focused attention on identification of the direct beneficiaries of development expenditure, including aid funds, with employment creation and the alleviation of poverty viewed as the main objectives. The 1969 DAC report went further, and identified maximisation of the spread of benefits as 'perhaps the most urgent issue facing the developing world'.[4] This

183

later evolved into an emphasis on 'basic human needs'. Since then, development agencies have maintained this position fairly consistently, although the 1978 DAC report declared that the concept of 'basic needs' had, perhaps, been oversold.

Numerous policy papers have been prepared discussing how to translate the overall objective of poverty alleviation into specific strategies. The choice of technology has been conspicuous among the issues covered in such papers, as having a bearing on the extent to which aid will achieve its stated objectives. At the same time, aid has come to be used more and more explicitly as an instrument to promote or perhaps even impose a view of the development process, so much so that Eckaus argued that aid agencies' 'leverage' was 'now virtually officially acknowledged in general though little, if anything is known in detail'.[5] A crucial question, then, is whether aid agencies are well equipped for this task.

Recipients have been relatively silent about the negative aspects of aid, for obvious reasons. A rather rare case of outspoken criticism is to be found in the following statement of Swaziland's Deputy Prime Minister at the 1980 session of the International Labour Conference:

> Inevitably, aid donors and international agencies have concentrated on giving help based on their own experiences. However, these experiences are not always relevant to Africa. They are derived from experiences of a system geared to high technology, large markets, high volume production runs, and high production costs. As a result, African countries are being equipped with goods and services which the developed world can most easily supply but which, regrettably, add more and more to the superstructure of African economies but are contributing very little to the base.[6]

Scholars have identified a number of problem cases. To quote a few relatively recent studies, Coulsen in Tanzania found it 'hard to think of more than a handful of projects financed by Western aid donors that are "successful" in the sense of providing the expected benefits at approximately the anticipated costs, without imposing significant unanticipated social costs on any part of a population'.[7] Randeni (1978) concludes his study of six industrial aid projects in Sri Lanka by pointing out, *inter alia*, the following facts: that the techniques adopted in these projects were clearly more capital intensive than known and locally available alternatives; that donors' influence was greater in 'package' projects covering various phases from pre-investment studies to actual operation than in non-package projects; that donors played an important role in choosing the (capital-intensive) techniques of construction of buildings; that projects established under aid were never subjected to rigorous cost–benefit analysis. The small domestic market and the absence of export market did not justify the large capital-intensive steel mill built with Russian aid: even after ten years of operation, the rolling mill could meet local demand by using a little over 30 per cent of the installed capacity. The fully mechanised brick and tile mill set up with a Rs. 2.2 million grant from the Federal Republic of Germany is portrayed as one of the most modern factories in operation in Sri Lanka. However, the only advantage of the most capital-intensive technology was a marginal improvement in quality. Although the project was a gift, the total amount of domestic resources diverted to it was about Rs 1.2 million. Without aid, less than Rs 0.5 million of domestic resources would have been adequate to establish two labour-intensive plants to produce the same output with three times as many workers.[8] With reference to agricultural mechanisation in Sri Lanka, Burch (1979) found a correlation between the levels of aid-induced tractor sales in this country and surplus capacity in the British tractor industry.

Frequently, aid projects have come under attack in the donor countries themselves. Indeed, the numerous critics of aid in donor countries have a marked tendency to concentrate on the so-called 'horror stories' – projects which for one reason or another, obviously 'failed' – as evidence that aid does not work. For example, Sweden's 'biggest aid project ever', the Bai Bang project on paper and pulp industry in Vietnam, has drawn considerable criticism from different quarters in Sweden: the project would be too big, too capital intensive, too expensive in terms of final product, too complicated to be implemented by development assistance authorities.[9] As we will note below, aid agencies often talk frankly about their 'failure cases'.

Of course, it is not always the donor that is to blame. In Randeni's study, the government was found to prefer capital-intensive techniques because of the availability of cheap capital. The political party in power attached greater importance to larger projects as symbols of their achievement. One reason for establishing these projects without adequate planning was the pressure exerted by the then ruling party to complete the project during its own term of rule. Commercial as against socio-economic considerations were usually overemphasised by national bureaucrats. Burch also argues that: 'In the case of Sri

Lanka, there were many groups – cultivators, politicians and bureaucrats – whose interests were served by the import of tractors, in different ways and at different times. And the process was reinforcing; once tractors were imported, the ownership, control and advocacy of them created new economic and institutional sources of support for the technological option they represented.'

In his essay on the choice of technology for irrigation tubewells in what was then East Pakistan, Thomas investigated why the World Bank and government officials concurred in a less than optimal technology. His conclusion is that, because of shortage of time and information, the Bank's appraisal mission chose the technology

> that seemed to fit their perception of what was needed, met their organisation requirements, and also had a satisfactory economic justification . . . As far as the government officials were concerned, aid was available only for medium-cost wells and this technology conformed with their institutional requirements and perhaps their personal preferences as well . . . In the actual decision making, such factors as risk avoidance, appearance of modernity, established procedures, familiar techniques, and by no means least, control, outweighed development policy objectives.[10]

Since external development finance may have a very considerable impact on the recipient economies, especially low-income ones, such criticisms cannot be neglected. Accordingly, the present study was designed to investigate the following questions:

(1) In what ways, and with what success, has the aid agencies' growing sensitivity to the importance of the choice of technology as a policy issue been translated into practice?
(2) In what ways, and to what extent, do the inherent biases of aid practices constrain or distort the choice of technology?
(3) How far and in what ways have the aid agencies been trying to cope with such criticisms as those contained in the above-cited studies?

The scope of the study is confined almost exclusively to financial aid explicitly addressed to the promotion of development. It is intended to examine the extent to which aid agencies' rules and practices are compatible with their stated objectives. A major deficiency of the study is that it was not possible to check conclusions in the form of reasonable suppositions, based on analysis of stated practices, against empirical evidence in the form of case studies. The analysis is based partly on the published literature, but mainly on a questionnaire-based survey of aid donors which was carried out in 1978 and 1979.[11] Twenty of them replied to the questionnaire:

Multilateral agencies: The World Bank; the African Development Bank (AfDB); The Arab Fund for Economic and Social Development (AFESD); the Asian Development Bank (AsDB); the Caribbean Development Bank (CDB); the Inter-American Development Bank (IDB); and the Commission of the European Communities (EC).

Agencies of DAC countries: Australia, Belgium, Canada, Denmark, Finland, Netherlands, New Zealand, Norway, Sweden, Switzerland, the United Kingdom, and the United States.

OPEC countries: Kuwait (the Kuwait Fund).

Centrally planned economies: Czechoslovakia.

After receipt of the responses, visits were paid to selected countries or agencies whose responses emphasised the significance of a number of new questions. These were Belgium, Canada, the United Kingdom and the IDB, as well as the Federal Republic of Germany (because of the concentration on industrial technology) and the Commission of the European Communities. The Netherlands had already co-operated in the pre-testing of the questionnaire. Very roughly, the agencies covered by the study accounted for about three-fifths of total DAC official development assistance (ODA) and about one-quarter of the developing countries' total receipts of external finance (including direct investment, export credits and bank loans).

TECHNOLOGY AS A POLICY ISSUE

The first step in the analysis was to establish the extent to which choice of technology figured as an issue in overall statements of general policy. But general policy statements – in terms of individual agencies' or governments' view of the role of aid in promoting development – were surprisingly difficult to find. If asked what was the current policy of an aid agency such as the World Bank or the regional banks, most observers would cite its president's speech at the latest annual meeting, an occasion commonly used for presenting broad strategic shifts. Yet no multilateral respondent in our survey referred to any such statement as reflecting current policy. The policy documents cited were in all cases statements of the operational rules according to which these agencies operate. In two cases, simply the articles of association were referred to. This seems to suggest that one should not attach too much significance to the kind of very general statements which tend to be widely quoted. At the sectoral level, one tends to find much more specific policy directives concerning what the agency should actually do.

The bilateral agencies of the DAC countries are, generally speaking, directly responsible to legislatures. Consequently, they seem to attach greater importance to general policy statements than the multilateral agencies. In the light of our findings, such statements fall roughly into three groups:

(1) general statements of the underlying philosophy of the aid programme (e.g., Federal Republic of Germany, Canada);[12]
(2) statements, often in the form of ministerial addresses to the legislature, which describe in broad terms the kind of relationship with recipients that is being sought (primarily Sweden and the Netherlands);[13]
(3) statements of specific guidelines for the aid programme (e.g., the United Kingdom and the United States).[14]

In general, these policy papers reflect individual agencies' current interpretation of the poverty-focused development strategy. They appear to have been influencing the geographical, rather than sectoral, distribution of aid, as testified by the rapid increase in the concentration of aid among low-income countries and especially the least developed countries in recent years.

Regarding technology as a policy issue, a distinction needs to be drawn between narrow and broad approaches. The first starts with the particular policy objective of promoting new technologies, and leads to the allocation of some (usually small) proportion of the aid budget to that purpose. The second emphasises the diversity and project-specific nature of the issues to be considered, and leads to the establishment of procedures that are intended to ensure that the choice of technology will be carefully considered on a case-by-case basis.

Elements of both approaches are found in the policies of most agencies. But there are differences of emphasis. Multilateral agencies strongly favour the broad approach. This is the case with the World Bank, in particular. A major review of appropriate technology in the

World Bank assisted projects in 1977 gives the following number of projects and volume of Bank/IDA lending, in which choice of technology had been an important issue:[15] rural development (26: $408.2 million); education (7: $98.3 million); urban development (5: $131.2 million); tourism (2: $42.0 million); highways (6: $161.5 million); total: 46 projects, $841.2 million. The IDB publishes a regular report on the incorporation of appropriate technology in its operations. The report for June–December 1977 covers irrigation, agricultural extension, agricultural credit, small-scale fishery, artisan industrial development, technological research and development, small industry credit, rural telegraphy, potable water, rural health services, low-cost housing and rural training.[16] Eighteen projects were involved, with a total of $236.3 million in IDB financing. Here, again, the importance of technological choice is recognised in a wide area of activity.

Programmes of bilateral aid donors explicitly aimed at appropriate technology were initially confined largely to research, and were implemented through an institute especially set up for this purpose. (Editors' note: the more recent activities of these institutes have been extended to include operational programmes, as indicated in their annual reports. For a survey of relevant evidence, see Jéquier and Blanc (1983).)

In Belgium, the 1977 budget earmarked Fr.B.50 million for research and dissemination of appropriate technology. Rapid disbursement of this allocation led to a further allocation of Fr.B.47 million in the 1978 budget. The main focus has been on alternative energy sources, especially solar energy, and the authorities have sought the co-operation of Belgian universities in setting up a research centre on alternative energy sources in Rwanda.

In Canada, support for appropriate technology falls in the field of competence of the International Development Research Centre (IDRC), where technology is regarded as the preserve of a few enthusiasts. CIDA itself, however, reported that it had recently developed an 'active interest' in alternative energy systems.

In France, the Groupe de recherche sur les techniques rurales appropriées au développement (GRET) was established at the instigation of the Foreign Ministry in 1976, and the Comité français des concours d'inventions et d'innovations adaptées aux régions en développement (CIARD) in 1977 with the aim of orienting French innovative capacity to the needs of developing countries.

In the Federal Republic of Germany, a technology 'clearing house', the Technology Transfer Co-ordination Centre (TTL) was set up to help identify the needs of developing countries regarding solar energy, food technology, health, agricultural engineering, etc. It was jointly financed by the Ministry for Research and Technology and the BMZ. In 1977, this was absorbed into the Gesellschaft für Technische Zusammenarbeit (GTZ) as the German Appropriate Technology Exchange (GATE). GATE officials replied to our enquiry that special effort was being made to prepare dossiers on appropriate technologies which included drawings. Of the eleven dossiers mentioned as already completed, eight were related to alternative energy sources.

In Japan, an Institute for the Transfer of Industrial Technology (ITIT) was established in 1973, with the aim of developing new technologies for developing countries or adapting existing technologies to local conditions. Financed by the Agency for Industrial Science and Technology, it draws on sixteen national institutes and laboratories.

In the Netherlands, a TOOL Foundation was set up in 1974, 'to provide a link between appropriate technological knowledge and experience and basic needs in the rural and intermediate sectors in the Third World'.[17] It is financed largely from the Foreign Ministry's aid budget.

In the United Kingdom, a working party established by the Ministry of Overseas Development initiated a programme in 1977–8 to promote, and disseminate information on, 'intermediate' technologies. The resources allocated are channelled primarily through the Intermediate Technology Development Group (ITDG).

In the United States, the International Food and Assistance Act, 1975, provided for the establishment of a private non-profit corporation, Appropriate Technology International (ATI). At the time of writing, ATI was still going through an extended formative stage. So were more ambitious programmes for which budgetary allocations had been made.

The 1978 DAC report expresses some concern that these research institutions might be obliged, by regulation, to offer patent rights for resulting technologies to their own authorities.[18] This is certainly the case in some instances, for example, the British Tropical Products Institute. To the extent that this concern is justified, then the narrow approach will become a formula for reinforcing the developed countries' dominance in advanced technology with a new dominance in technologies specially designed for developing countries.

From the above, both multilateral and bilateral aid agencies ap-

pear to have been incorporating the concept of appropriate or intermediate technology into their programmes, to a smaller or greater
extent, but with a very sharp distinction between those agencies
which see technology as a pervasive issue and those which see it as a
subsidiary aspect of policy, with a specific but usually not large
budgetary allocation. The broader approach is clearly in many ways
the more convincing, but much more difficult to put into practice. In
the narrow approach, officials need only to consider projects in which
choice of technology is clearly identified as the dominant issue. The
broader approach requires the setting up of procedures which will
ensure that the technological aspects of all projects are carefully
considered, alongside all the other criteria – often conflicting criteria
– that need to be taken into account.

AID AGENCIES' INFLUENCE ON THE PROJECT DESIGN

A large majority of the officials interviewed in this study felt that
their influence on the choice of technology was slight. From a typical
desk officer's viewpoint, an agency enters the project cycle (i.e., the
process of identifying, preparing, appraising, modifying, approving
and implementing projects) when it receives a project proposal. Most
agencies take the view that by then the basic questions of project
design have already been settled, and that they can hardly propose
substantive modifications, partly because of lack of information, but
more because of the political constraints of an arm's length relationship. The only way to have effective influence would be to enter the
project cycle at a much earlier point. However, recipients might take
that as an illegitimate degree of intervention. Officials who took this
self-restraining view tended to argue, not only that the influence of
aid was marginal, but that it would be unreasonable to expect it to be
otherwise.

All agencies, of course, retain the power to refuse a project.
Officials of some agencies reported that they preferred to reject
clearly ill-designed proposals out of hand, rather than suggest major
modifications. Some others, however, argued that even this power
was relatively ineffective, unless the recipient was exceptionally
dependent on a single donor or donors were extremely well co-
ordinated. Most recipients could usually find alternative sources of
finance and they were becoming increasingly adept at doing so. Also,
it was argued that agencies found themselves already committed in

principle to projects by the time they came to the stage of appraisal, and that it was then difficult to withdraw.

Indeed, if aid agencies had been rejecting what they considered ill-designed projects, not so many mistakes might have slipped through. Few agencies would deny the existence of mistakes, and some failure cases have been acknowledged more or less publicly: sugar in the Ivory Coast in the case of Belgium, a bakery in Tanzania in the case of Canada, pulp and paper in Tanzania in the case of the United Kingdom. (Multilateral agencies are less willing to acknowledge such mistakes.) A recurrent feature of these acknowledged mistakes is that large-scale advanced technology was used, requiring managerial capacity greater than was locally available and possibly displacing smaller-scale more labour-intensive alternatives. Questioned about specific causes of such mistakes, officials often referred to pressure from the recipient government and pressure from donor countries' export-oriented capital goods industries suffering from excess capacity. Others contented themselves with a wry smile and remarks like, 'We are learning all the time.' Aid agencies sometimes do not realise the real cause of the problem of 'inappropriate technology' choice. To some extent, this may be attributable to their lack of experience in the recipient countries as well as their highly oversimplified view of the choice of technology (e.g., straightforward factor substitution). Sometimes, aid agencies fail to persuade recipients to modify their project proposals in a more appropriate direction simply because of tactlessness: one of the agencies interviewed had agreed to finance a small-industry project in Pakistan and then tried to persuade the local authorities to install Indian equipment!

However, aid agencies can influence project design unintentionally. First, recipients are likely to choose and design projects largely with aid agencies' preferences in mind. This is the standard refutation of the theoretically correct argument that resources are fungible and that aid agencies are in practice financing marginal projects, rather than the projects to which their resources are notionally attached.[19] Second, even the most highly project-oriented of agencies, such as the World Bank or the BMZ, seldom finance projects in isolation. The point of entry into the project cycle is also the point of entry into a process of sectoral development. For example, an Afghan request for financing grain storage led to a review of the marketing mechanism and price stabilisation scheme, and hence to a series of proposals for the construction of an integrated network of transport and storage facilities. While the influence of an agency on the first project may be slight, the impact of its subsequent decisions on how it follows up that initial project may be considerable.

Some agencies, in an effort to move away from the project-by-project approach, have developed procedures for sectoral planning within a framework of country programming. Other agencies appear to be reluctant to become so deeply involved in the recipients' internal processes. Most agencies insist that in formal terms a request should come from the recipient government: and among the questions that figure in agencies' checklists one of the commonest is, 'Is the project in the recipient's development plan?' However, there are several ways in which agencies help recipients to firm up their proposals. For instance, the foreign minister of a donor country, making a tour of a region in which his government wants to enlarge its presence, is likely to let it be known through embassies that project proposals will be welcome, and likely to receive a positive response. Almost all technicians in aid agencies regard this approach with distaste. It saddles them with a political commitment to projects which are not open to technical appraisal, and which are quite likely to end up as prestige projects, to be seized upon by their own domestic aid lobbies as examples of the agency's technical incompetence. Yet it is difficult to see any great difference between this approach and the professionally more respectable device of a country programming mission, which is expected to come up with a list of well-prepared project proposals at the end of a few weeks. Most agencies take the view that once a commitment in principle has been given the scope for influence is negligible. Yet there are strong arguments in favour of entering rapidly into such a commitment, so that discussion is not distorted by the posturing that recipients feel compelled to take in order to attract funds.

Some agencies both acknowledge this dilemma and believe that they have, to some extent, resolved it: SIDA, British ODA, the USAID, the IDB, and the EC Commission. These agencies all have relatively strong resident missions, which provide the day-to-day underpinning of a programming process. That is clearly a necessary condition, though not a sufficient condition, of effective participation in the process of project design. If officials are right in thinking that influence on design is ineffectual once a proposal has been submitted, and if such influence is thought desirable, delegation of powers to overseas missions appears to be the direction in which to move.

In addition to these visible instruments for the stimulation of project proposals, several covert instruments are used to make the recipient appear to be asking for what the donor wants to provide. This is known as 'ventriloquising'. At its crudest, a donor will say, in effect, 'If you ask for our aid for Project A, we shall say "yes"; but if you ask for our aid for Project B, we shall say "no".' In practice, this appears to have become rare. A large majority of officials interviewed forcefully denied that they were in a position to stimulate project proposals so directly. At the same time, many of them conceded that proposals were frequently stimulated by processes which they had themselves set in train.

The ventriloquiser's interests can be commercial, personal or institutional.

In commercial ventriloquism, the proposal usually emanates from the overseas representative of an equipment manufacturer, whose main aim is to sell as much equipment as possible. Commercial ventriloquism, therefore, is by and large restricted to bilateral programmes, since the procedures of multilateral agencies go a long way towards precluding commercial representatives from reaping such gains. However, it is not restricted to tied aid, since the principal strength of the commercial representative's position is not the likelihood of a formal procurement restriction, but the possession of advance information derived from informal contact. Among the agencies examined in detail, awareness of commercial ventriloquism was sharpest in the case of the Federal Republic of Germany, which provides mainly untied aid!

In aid flows, personal interests come to the fore primarily through technical co-operation. Several officials interviewed expressed concern over the propensity of technical assistance personnel to steer recipient governments in the direction of equipment suppliers known to them in their own countries. The propensity is not necessarily corrupt. In good faith, a consultant is likely to mention the suppliers he knows, which may give rise to a request for financial aid the outcome of which is virtually predetermined. The problem is widely recognised, and remains unsolved. The obvious solution lies in the improvement of information flows.

Institutional interests appear strongest among those agencies which believe that the principal constraint on increased aid flows is a lack of absorptive capacity, or aid-worthy projects. The agency's natural response is to put a special effort into project identification, and with the best will in the world such an effort tends to result in project proposals moulded in the agency's own image. At the extreme, one may end up with project identification missions to stimulate proposals of the kind that the agency wants, project preparation missions to ensure that they are in the form that the agency wants, and project appraisal missions to add a gloss of perceived influence – a serial relationship which Mason and Asher[20] describe as 'incestuous'.

The foregoing should suffice to show that the self-portrait of aid agencies' loan officers passively waiting for proposals and constrained by respect for recipients' views, possibly misguided, of what they want, is not one that bears very close examination. Strengthening of country programming procedures is constrained by legitimate concerns for national sovereignty. Codification of the hidden procedures for ventriloquising requests is inhibited by the faintly disreputable air that hangs about such practices, blocking explicit and systematic discussion of how to make them more effective. Only when a proposal has been formally received can anything approaching a rule book come formally into operation. What happens before that, with both positive and negative effects on the choice of technology, is likely to remain relatively uncontrolled.

PROJECT SELECTION AND PROCUREMENT

Project Appraisal Methodology

Since the late 1960s, sharpened by the work of Little and Mirrlees,[21] there has been a growing literature on project appraisal. The earlier works were focused on the elimination of factor price distortions. More recent writers, notably Squire and van der Tak,[22] have argued that the use of any measure entails a value judgement and that the value judgements should be made explicit, e.g., with poverty weightings. In either case, use of a proposed methodology should lead to the identification of labour-intensive alternatives, for the benefit of the poor.

Administrators commonly complain that the new methodology is too complex and sophisticated. Furthermore, as soon as weightings based on value judgements are explicitly introduced, it becomes relatively easy for economists in recipient ministries to cook up figures to produce the results that their various audiences require. Nevertheless, some aid agencies have tried to get back to basics. The British ODA, for instance, has a project manual which incorporates a simplified version of the methodology. Others play the new methodology for what it is worth, without allowing it to influence their decision-taking processes in any clearly identifiable way. The World Bank, for example, having sponsored the work of Squire and van der Tak, has yet to demonstrate that this has significantly influenced project decisions, and with what results.

A number of agencies simply list a set of criteria by which projects should be appraised. This approach is vulnerable. For instance, an agency very much in favour of 'appropriate technology' is likely to feel hopeful about a project proposal with that label, without carefully studying whether the 'appropriate' technology is indeed appropriate. Some agencies have detailed checklists. If the list is very long, a question arises concerning the weights attached to individual questions. Even if choice of technology is among the listed questions, it may be included as no more than something to be ticked off.

Assessment of the weight to be given to each question in a checklist becomes easier if separate lists are prepared for different sectors. The United Kingdom has sectoral checklists as well as a project manual. The checklists contain questions about the choice of technology, but not in a form which reflects the direct influence of recent policy guidance in the manual. Interviewed officials attributed this to the fact that the two kinds of documents had been prepared by different groups of the Ministry's staff.

The question of sectoral checklists leads on to a broader question about sectoral policy papers. In the case of multilateral agencies, sectoral policy papers appear to be regarded as the core of operational guidance, and this propensity seems to be spreading to bilateral agencies. In CIDA, for example, considerable effort was being put into the development of new sectoral policies, at the time of writing. In the Federal Republic of Germany an interesting attempt has been made to identify the effective range of technological choice to be considered in various sectors, arriving broadly at the following conclusions:[23]

(1) considerable flexibility in agriculture, with variations in the appropriate level of mechanisation from country to country and from region to region;
(2) interchangeability of factors of production in infrastructure, especially in roads and dams;
(3) ranges of choice determined primarily by relative costs in services (education, public health, transport);
(4) in manufacturing, considerable variation between industries, with maximum flexibility in clothing and foodstuffs, and considerable flexibility in wood, paper and rubber.

This may look crude, but this is the kind of differentiation which is easiest to translate directly into policy. However, the paper in question is not a policy paper. In almost any aid agency, some people, often those in central positions, have given quite precise thought to questions concerning the choice of technology in different policy areas. It is rarer, though, to find policy papers that reflect that same precision. Moreover, sectoral policy papers often fail to provide clear guidance. The World Bank's sectoral policy paper on development finance institutions,[24] for instance, emphasises both profitability and employment creation, without settling the question whether these two requirements are in conflict and in what ways the conflict should be resolved.

Interpretation of the Rules

How the existing rules are interpreted and how discretionary powers are exercised can have a crucial influence on an aid agency's performance. This may be best illustrated with reference to the EC. On paper, its procedures appear to be significantly ahead of most other agencies in bringing practice into line with current policy perceptions:

(1) substantial participation by the recipient in the programming process;
(2) multi-year commitment, which permits forward planning on the basis of known aid availability;
(3) a major delegation of authority to field missions, not only for project identification, but more importantly for implementation;
(4) indifference between local and foreign costs in determining eligibility for finance;
(5) willingness to subdivide contracts, which leaves considerable scope for local enterprise;
(6) adequate price discrimination in favour of local suppliers and contractors;
(7) clear specification of the criteria for procurement from alternative suppliers in non-member countries;
(8) formal provision for accelerated and simplified procedures for small- and medium-scale projects;
(9) explicit adaptation of established procedures in sectors, notably agriculture, in which procedures developed originally for conventional infrastructure projects are likely to be inappropriate;
(10) a stated willingness to modify procedures for joint financing of projects.

Yet recipients, particularly anglophone countries which are unfamiliar with French traditions, feel that the Commission is extremely inflexible and unwilling to take account of the special features of individual cases. This problem seems to be attributable mainly to the following factors. First, the Commission's discretionary powers are applicable only 'exceptionally and on a case-by-case basis'. Second, its operations are monitored exceptionally closely by member countries, which are quick to complain if they think that another member has been given an unfair preference. In the course of the negotiations for the second Lomé Convention in 1979, no fewer than forty-four areas were identified where implementation could be improved if the Commission were to take a more liberal view of the powers it already possessed.[25]

If one of the most progressive aid agencies, in modifying practice in the light of current policy perceptions, is regarded as one of the most restrictive by recipients, one realises that modifications to the rules are not enough. It is no less important to know how these modifications are to be interpreted.

Conflicting Objectives

An aid agency is likely to be pursuing multiple objectives, some of which may well be in conflict with each other. The possibility of conflict with the objective of ensuring an appropriate choice of technology is eliminated only if technology is regarded as the overriding issue. Few agencies would accept such an approach. Commercial objectives are usually cited as the leading contender, though the tendency to see aid primarily as an instrument of foreign policy may still persist in some cases. In practice, those agencies which are most clearly exposed to commercial pressures seem to run two types of programmes, individual projects being more or less explicitly classified as 'developmental' and 'commercial'. The Federal Republic of Germany clearly and openly distinguishes poverty-focused development aid, mainly channelled to poor countries, from more commercially oriented high-technology aid, mainly intended for middle-income countries. In Canada, contributions to multilateral agencies are controlled by the Ministry of Finance, while projects involving major sales of Canadian hardware are subject to examination by a body which has strong participation by the Ministry of Industry, Trade and Commerce.

This difference may imply a variety of non-developmental objectives. In many countries, certain portions of aid are left out of the aid agency's control altogether, to give the government a free hand in providing, e.g., politically motivated aid.

Commercial pressures seem to have intensified since the beginning of the recession in 1973, and successive DAC reports exhibit increasing concern over this issue. Aid administrators in several countries mentioned the existence of underutilised capacity in an industry as a major factor which could drive a proposal through, regardless of the fact that the notion of underutilised capacity as a valid guideline for aid allocation was attacked and generally discredited even in the 1960s.

A serious conflict can also arise between the programme objectives to be achieved by the aid project and the managerial objectives of the agency's administrators. The latter objectives are likely to be more conspicuous in multilateral agencies, because of the greater degree of autonomy they enjoy. Expansionism, for instance, leads to higher commitment or disbursement targets, which in turn tend to favour large and familiar projects, under the pressure to 'move money', and discourage a lengthy search for and careful scrutiny of technological alternatives.[26] This was indeed cited as *the* dominant consideration of the interviewed officials at one of the major agencies, which has been at the forefront of the campaign for a more poverty-oriented development strategy. To the extent that 'appropriate technology' is equated with small-scale technology, it might cynically be argued that one way to encourage its application would be to cut the flow of resources, forcing a search for administration-intensive activities which could keep the same number of staff employed on a reduced budget. As a matter of fact, a markedly sharpened focus on appropriate technology at one of the bilateral agencies examined seemed to be closely associated with a recent across-the-board cut in the aid programme.

Practice of Tying Aid

Among the inherent biases of aid, the one which perhaps most obviously affects the choice of technology is the capital- and import-intensive bias which arises from the practice of tying aid to the import content of projects. Procurement tying, that is, the restriction of procurement to goods and services supplied by the donor country, raises similar problems.[27] However, the volume of aid which is formally tied is now relatively small as illustrated by the following figures for gross disbursements in 1977 (in million dollars):[28]

	Untied	Partially tied	Tied	Total
Multilateral	3 529.7	875.2	226.2	4 631.1
DAC bilateral	4 327.4	1 432.8	6 270.1	12 030.3
Total	7 857.1	2 308.0	6 496.3	16 661.4

The figure of $6270.1 million for tied bilateral aid includes some $4000 million of food aid and technical assistance. So, the proportion of tied aid for purchases of hardware was already relatively insignificant in 1977, and it has continued to decline since.

At the same time, a number of devices have been introduced to mitigate the problem of tied aid. Most agencies which give tied aid waive the tying clause if a strong case can be made. However, these provisions appear to be rather underutilised. Canada, for instance, specifies a fixed proportion of the overall programme that can be used for third-country procurement. That ceiling has never been reached. One possible explanation is that recipients approach Canada with Canadian equipment in mind. Another possible explanation is that recipients overestimate the extent to which 'tied' aid is really tied and do not know what exceptions are possible and under what conditions. To make a case for a waiver one needs a very precise knowledge of the rules which, in some cases, are confidential. Also, if the rules are complex, as they usually are, their application may be discouraged by the prospect of extended negotiation and delayed project implementation.

There is a growing tendency to combine procurement-tied aid with an element of local cost financing. For example, the United Kingdom sets country rather than project limits on the proportion of funds that may be used for local costs, which, in principle, should give considerable flexibility within individual projects. A more recent variant is 'partial untying', whereby aid is made available for goods and services from the donor country or from any developing country. As with third-country waivers, the facility seems to have been underutilised. Officials of the relevant aid agencies have formed an impression that developing countries have a strong bias against buying from other developing countries.[29]

ORGANISATIONAL PROBLEMS

Project Staff

By and large, agencies incorporate sectoral and other technical project skills into their organisational structure in the form of either an operational 'projects' department or a small unit of specialist 'advisers'. The first is the structure normally adopted by multilateral agencies. In the case of the World Bank, the growth of its programme rendered this structure unwieldy, and project staff were integrated into the geographical departments. This appears to have facilitated closer co-operation between generalist loan officers and technical specialists. The regional banks have never had the problem of familiarising technical specialists with local conditions, and have on the whole retained the traditional structure.

The smaller size of the bilateral agencies does not justify such an arrangement, and they try to incorporate professional perceptions early in the project cycle through a central advisory staff. Canada and the United Kingdom are conspicuous examples. In these countries, the senior personnel concerned appear to be both influential and alert to issues of technological choice, though rather sceptical of certain concepts of 'appropriate technology'. In the latter country, geographical loan officers emphasise that reference to the engineering advisers is the first step that they will take virtually automatically in assessing a project proposal. However, such advisers are exclusively civil engineers, specialising mainly in water and roads, reflecting the sectoral emphasis of the British aid programme. Advisers in other fields technically enjoy the same status, but tend to be further removed in semi-autonomous units such as the Tropical Products Institute.

It is interesting to note that in both countries the engineers gave more weight to cost effectiveness in the choice of technology than authors of much of the recent literature on the subject thought they would.[30]

Agencies without in-house expertise resort to various makeshifts. Of these, the commonest is reference to specialists in other ministries. Compared with in-house expertise, advice from other ministries is likely to come rather late in the project cycle. It also tends to carry less weight in internal decision-making. Moreover, officials in other ministries may be less aware of current thinking on development problems and more influenced by their own domestic experience. In fact, none of the attempts to improve the technological aspect of project design identified in our survey originated from such 'other' ministries or government agencies.

Field Missions

A theme that runs through much of the aid literature concerns the superficial nature of the judgements made by peripatetic experts. A visiting specialist, even a full-scale visiting mission, is unlikely to acquire much feeling for local constraints, compulsions and other factors relevant to project design. In this respect, the advantages of maintaining resident missions are many and obvious. In the context of the choice of technology, and particularly of small-scale projects, for example, they can make a more realistic assessment of available managerial capacity and other relevant resources. Two possible disadvantages are that their presence can be seen as foreign intrusion and that they can lose touch with the lessons of experience learnt elsewhere. So far as the bilateral agencies are concerned, the first of these dangers has largely been taken care of by the diminution of their size. The second danger can be easily averted by supplementing field missions with visiting specialists, by staggered rotation of their staff, or, as in the British case, by establishing regional rather than country missions.

Of the multilateral aid agencies, only the IDB has had a strong tradition of delegating responsibilities to country missions. The EC started with country representatives. Under the Lomé Convention and the Fourth European Development Fund there was a rapid build-up of resident missions. Delegation of powers, however, appears to be confined largely to the implementation phase, about three-quarters of contracts being awarded through field missions. Project approval still appears heavily centralised in Brussels. Consequently, one observer felt that approval of projects took longer compared with other donors.

Among DAC bilateral donors, France has the strongest tradition of resident missions. Among the respondents to our questionnaire, Denmark, Sweden, the United Kingdom and the United States maintained missions. The operations of the resident missions of the EC, Sweden and, more loosely, the United Kingdom are set in the framework of a multi-year programming exercise.

In the light of growing recognition of the relevance of local conditions to technological choice, the question of field representation deserves serious study. Since a multiplicity of resident aid missions might cause dismay among recipients, this may also need some investigation.

Executive Agencies

Multilateral aid agencies are relatively self-contained. Bilateral agencies, however, frequently work through a variety of executive agencies, of which three main types may be distinguished: (i) implementing agencies, which undertake certain delegated tasks under instruction from the aid agency; (ii) semi-autonomous agencies, which have delegated power to use their own discretion in deploying the funds allotted to them by the aid agency in certain defined fields; and (iii) advisory agencies. Implementing agencies may be divided into agencies primarily concerned with project appraisal, possibly including the conclusion of loan agreements, and agencies primarily concerned with implementation, including loan administration and procurement.

A conspicuous example of an implementing agency of the first type is the Kreditanstalt für Wiederaufbau (KfW) of the Federal Republic of Germany. The BMZ does not make detailed technical appraisal of project proposals, but passes them to the KfW for comment. The KfW, however, considers that the proposals passed to it by the BMZ have already been approved in principle, and that its own function is at best to propose minor modifications, and at worst to rationalise decisions already taken at a more political level. This implies that at least some ill-designed proposals slip through.

The KfW's consciousness of its independent status, reinforced perhaps by a justifiable pride in its very high reputation as a banking institution, makes it resistant to the imposition of the BMZ's criteria on its own technical judgements. KfW officials were very firm in their view that labour intensity was not an issue in the KfW's assessments, and that the concept of 'basic needs' lay far outside their institutional perspective which was focused on 'strict banking criteria'. In addition to loans administered on behalf of the BMZ, the KfW in its own right provides official export credits to developing countries, to support the country's high technology industries. At the same time, its reputation is well established in supporting the launching and growth of local development finance institutions. Asked why the financing of small-scale industries appeared to figure only as a very minor element in KfW support for local intermediaries, the officials in charge replied that the local development finance institutions themselves preferred medium-scale industries, mainly because less risk was involved.

Aid agencies tend to regard a procurement agency, such as the British Crown Agents, as a purely executive body which helps reduce the amount of their own paperwork. However, a procurement agency may have a more positive value than that, especially if staffed with professional engineers. As is the case with the Crown Agents, for example, it may become deeply involved in procurement tasks that are normally left to the recipient, so that it may identify design problems that would otherwise not be brought to the aid agency's attention. In this respect, its advantage lies in its thorough knowledge of the aid agency's rule-book (e.g., the limits on waivers regarding tied aid).

Among semi-autonomous agencies, the most important for the purpose of this study are the development corporations established by several European donors – the Commonwealth Development Corporation (CDC), the Deutsche Entwicklungsgesellschaft, the Danish Industrialisation Fund for Developing Countries, the Netherlands Finance Company for Developing Countries (FMO), and others. Having a somewhat more commercial orientation than aid agencies, though still broadly within a developmental framework, these corporations tend to see the choice of technology rather strongly in terms of cost effectiveness, reliability, and other factors which affect profitability. This does not necessarily militate against the concerns of those who view the choice of technology from the standpoint of development objectives. A relevant question, however, is to what extent these corporations employ technical specialists who can adequately assess the capability of technical partners and the technological alternatives that deserve consideration. One view advanced is that to employ a full range of specialists merely duplicates the skills and experience for which technical partners are sought, while to employ merely one or two general engineers is hardly worthwhile, since they can always be out-argued by specialists in particular industries. According to that view, the solution is to build up a relationship of trust over time, and to avoid further co-operation with companies whose performance record has proved unsatisfactory. With its direct and extensive involvement in management, the CDC appears to have found a more reliable solution. Its involvement in management overseas also gives it the advantage of extensive knowledge of applied research carried out in developing countries (e.g., crop research), which it appears to find more useful than comparable work carried out by technological research institutes in the developed countries. Clearly, anyone concerned with the use of aid to promote more discriminating application of technology in developing countries should give these corporations close attention. In the United Kingdom, this view has been officially endorsed in the relevant policy paper.

Research institutes can be a main source of advice for aid agencies. Where their relationship is formalised, it seems to work quite well. This is the case, for instance, with the Tropical Products Institute, which is technically a part of the Overseas Development Administration and whose senior staff have official status as advisers to this institution. A larger number of other research institutes are 'independent' on paper but depend heavily on an aid agency for their research funds and in some cases their regular budget. Most of the institutes specialised in appropriate or intermediate technology are of this type. One may also mention certain institutes of development studies. Officials of aid agencies interviewed, however, reported that they rarely sought advice from these institutes. To some extent, this seems to be due to the usual language problem between researchers and decision-makers. During our survey, it was also discovered that, out of fear of interference in their research programmes, the research institutes often preferred an arm's-length relationship. In one case, it was stated very firmly that it was not the institute's task to advise the aid agency.

Consultants

The literature on the use of consultants in aid programmes, especially for the evaluation of technical assistance, is largely addressed to the question: Was their advice accepted? A prior question, which is less often asked, is: Did they give the right advice?[31] That is much more difficult to answer, since consequences of accepting alternative pieces of advice remain buried in the hypothetical realm. In spite of all the effort made, there is still no widely accepted and reliable method to evaluate their services.

The main protection that aid agencies have against the danger of appointing wrong consultants is in the recruitment process. Some agencies have recruitment methods which border on the casual 'old-boy network'. Others follow more formalised methods. The base of a formalised recruitment process is usually a registration system which starts with the distribution of a questionnaire, reissued at regular intervals in some cases. Such questionnaires are focused on consultants' assessment of their own capabilities, which naturally tend to be overstated. Agencies where in-house engineers are influential may review their replies in great detail.

Recruitment from the register normally takes the form of advertisement or invited tender. At this point the aid agency is still in a relatively passive position. Active involvement usually does not begin until an award has been made and the point has been reached where terms of reference can be discussed in detail. Widely varied views were given, even within one agency, concerning the extent to which the terms of reference were or should be influenced by the consultants themselves. Consultants like to be involved in drawing up the terms of reference, on the valid ground that it is a part of the process of problem identification for which their expertise is required. However, consultants may define the problem on the basis of their own experience, which may or may not be relevant. The danger is greatest where the agency lacks in-house expertise and feels obliged to rely on the consultant's view. Conversely, agencies in which engineers are strongly represented tend to emphasise that preparation of the terms of reference is their exclusive responsibility; and yet these are the agencies which are best placed to gain from an open discussion with the consultants. In the case of an agency with resident missions, project proposals can be examined on the spot before a consultant is appointed. Consequently, it probably has a fuller view of the problem than the consultant.

Ex-post evaluation of the consultants is weak in most cases. If their performance is thought to have been disastrous, that will be recorded, but such cases are relatively rare. The consultants will write their own reports explaining how their best efforts were frustrated by inadequate terms of reference, political pressures, lack of support facilities, recipients' failure to appoint counterparts, etc. The main information on the aid agency's file remains in most cases the consultant's original self-assessment. Other information, of course, is available to the aid agency's specialists. Consultants become known as good or bad performers, or as strong or weak in particular fields, and are used accordingly. Where this kind of information is combined with the register, discriminating choice of consultants may be facilitated. This may be easier in countries such as the ex-colonial powers, where there is a corpus of consultants consisting of specialists in developing country problems. The situation may be more serious in countries which do not have such specialists and rely on a broader range of consultants whose main business is domestic, and who regard aid agency contracts as a supplementary source of income.

An effort to devise a more accurately aimed recruitment system might eventually lead to something similar to that of the IDB. The relevant office maintains background information on some 5500 consulting firms. The files are kept up to date by continuing correspondence, rather than by questionnaires, at an estimated rate of 3000 exchanges a year. Performance appraisal is said to be systematic and continuous, with approximately 1400 performance appraisals a year. A borrower must present a list of three to six firms from which he wants to invite proposals. If a suggested firm is not registered, it can be registered, but such firms are usually dropped from the list to save time. In a highly specialised field, the bank is also willing to suggest a list of ten to twenty possible firms to the borrower. The list is submitted to a high-level selection committee. This is one of the oldest and most influential standing committees of the bank, no recommendation from which has ever been reversed. Once the list has been approved, the bank will make its own selection on the borrower's behalf, if the borrower does not have the capacity to do so; but in the great majority of cases selection is the borrower's responsibility.

The point needs to be made that the IDB's system depended crucially on the fact that the official in charge was a person of standing in the bank, who had been in the job for some fifteen years.

Procurement

A major organisational problem in procurement, especially in bilateral agencies, is that the procurement unit, which may in some cases be a separate agency as already discussed, seldom has discretionary power, and tends to see its own role as exclusively administrative. Difficulties in project implementation often arise in connection with the supply of equipment inadequate to local conditions. The procurement unit is rarely qualified to make a technical judgement regarding requirements for modifications or for supplementary equipment. Some aid agencies establish a project monitoring group which meets regularly to check progress in project implementation, but it seems a cumbersome way of resolving a relatively simple problem.

Ex post Evaluation

Ex post evaluation of a project is difficult for two reasons. First, compared with *ex ante* project appraisal, the methodology is still primitive, being confined in many cases to little more than the question, 'Did it work?' Second, it raises sensitive issues of internal relations. Nevertheless, the major multilateral agencies have recently made significant efforts to improve their *ex post* evaluation machinery, of which the best known is the World Bank's operations evaluation unit (DOE).

Few bilateral agencies have comparable evaluation machinery. At the time of writing, the United Kingdom was in the process of developing a system, and had already established inter-disciplinary project management teams, initiated by the geographical departments, which reviewed selected projects on a semi-formal basis. In addition, a project-monitoring unit was established.

An interesting variant is found in Sweden. Swedish *ex ante* project appraisal is highly simplified and focused mainly on assessment of the benefits expected from the project (e.g., employment-creating effects). Provision for *ex post* evaluation is included, and for this purpose independent consultants may be hired. The system has the advantage of reducing administrative costs and of providing an instrument for nearly automatic identification of problems that have arisen in practice. However, when a very large project such as the Bai Bang pulp and paper project in Vietnam goes wrong, the wisdom of not asking detailed questions in advance may be questioned.

SUMMARY AND CONCLUSIONS

This study was addressed primarily to the single question: How do aid practices influence the choice of technology? The answer is that we do not know, because aid agencies themselves do not know and on the whole seem not to have tried to find out. We tried to analyse the extent to which policies and procedures had been modified in the light of growing recognition of technology as a policy issue, and the steps that had been taken to put new policies into effect, and to eliminate some familiar and inherent biases arising from traditional aid practices. Our findings may help formulate some tentative hypotheses about the *likely* effects of these practices. A set of carefully structured case studies with a comparable framework is the next task.

The selection of projects for such case studies as exist appears to have been strongly biased by the purpose for which such studies were designed. Typically, they are focused either on projects which had become conspicuous because something obviously went wrong or on projects selected by aid agencies as 'success stories'. In neither case would it be legitimate to draw general conclusions about the impact of aid on the choice of technology. To determine cause and effect with any confidence, the approach would have to be refined in one or both of the following directions. A more reliable, but also more difficult, approach would be to identify a control group of comparable projects not financed by aid. A simpler alternative would be a historical approach, aimed at identifying the sequence in which technological questions were pre-empted, to be set alongside the sequence in which actions stimulated by the intervention of aid agencies entered the project cycle. In view of the difficulty of determining what constitutes a 'comparable' control group, the historical approach may be preferable.

Procurement restrictions have been central to past discussion of inherent biases in aid. For that reason, they were dealt with in earlier sections of this study. Certain efforts to overcome these biases have been noted: greater flexibility on local and recurrent costs, provision under tied aid for third-country procurement in special cases, accelerated procedures for small projects, etc. In addition, it may be noted again that a number of DAC donors have subscribed to the DAC resolution on partial untying, permitting procurement from other developing countries. The principal sources of bias now appear to be the managerial pressure to meet commitment targets, and insistence on international competitive bidding. Here, too, however, there are some signs of improvement.

In spite of all these efforts and improvements, it is possible to argue that certain types of project with which the concept of appropriate technology has come to be most closely associated in the public mind, though desirable in themselves, are not suited to aid financing, and that development agencies should therefore avoid them. This is argued on four different grounds:

(1) that they constitute activity in highly sensitive areas, which raise complex social and political issues which development agencies, operating at arm's length, are unlikely to grasp;

(2) that they depend crucially on local initiatives which the rather heavy procedures of development agencies and the foreign label attached to an externally financed project could weaken;

(3) that development agencies have rather limited effective control over project design, so that they should avoid projects which raise serious design questions and should spend less time trying, ineffectually, to promote a particular view of technology;

(4) that appropriate technologies are already much more readily available and more widespread in developing countries than is generally supposed, and that development agencies are less competent to assess these than people on the spot, so that they should stick to projects for which the developed countries' technologies, possibly with some degree of adaptation, are clearly appropriate, and for which the development agencies have something to offer which developing countries lack.

The first two of these have some substance, but there are some obvious solutions. One is a greatly increased reliance on, and delegation of authority to, local institutions, which is broadly the Swedish solution. If that is rejected, on the ground that local institutions lack the necessary competence (which is a weak argument, incidentally, because the build-up of competence is part of the process which aid is intended to support), an alternative is to rely more heavily on non-governmental organisations, which is a solution adopted by several agencies, notably USAID and CIDA. However, non-governmental agencies have their own particular approach, which is only suited to certain types of project. Therefore, some kind of categorisation of aid projects may be necessary. For example, the Canadians distinguish between (a) grass-roots projects, mainly administered by non-governmental organisations, (b) export-oriented projects in which sales or gifts of hardware are the main element, and (c) projects which are decided at the political level. Such a blunt distinction may be unacceptable to some people but it seems more sensible and realistic than trying to load all these mixed motives and conflicting interests on to all projects at the same time.

The argument that agencies have very little influence over project design was widely cited by officials in the development agencies themselves. Clearly, their influence is limited, but the claim that they have no influence at all is not plausible. Aid agencies do have a comparative advantage in certain areas. For example, they clearly have a comparative advantage in aid for the development of export industries, and in minerals exploration. To extend this list would be a useful collective exercise.

Beyond the collective exercise, however, attention needs to be given to the special capabilities of individual agencies. For example, the Federal Republic of Germany concentrates its effort on industrial projects and support for development finance institutions. The fact that this is clearly in its own interest is not in itself strong ground for criticism. No country would give vigorous support to a policy against its own interest. Its approach is only open to criticism if it can be shown that the country's interests are in conflict with, and override those of, the recipients. In fact, there is strong evidence that recipients approach the Federal Republic of Germany because they want the kinds of equipment that the country is prepared to offer, and that if they want something else they go elsewhere for it. That seems quite a healthy relationship, in which the onus is on the recipient to decide what he wants and where to seek it. Another country which seems to be moving in much the same direction is Canada, which recently started drawing up an inventory of Canadian resources which were relevant to developing countries' needs. The eventual outcome of that exercise deserves further study.

The question arises, then: can this approach be put on a more systematic and widespread basis? That is a question which must ultimately be addressed to the DAC. But it may be useful to consult heads of departments responsible for aid negotiations in the finance ministries or planning offices of developing countries to seek their views. It may also be necessary to involve the World Bank, which, as the world's largest and most experienced agency in procurement for developing countries, probably has more knowledge than anyone else of what is available from where, even though its own procurement rules prevent it from making full use of that knowledge. It would be helpful if the resultant inventory could be published, unless there is good reason to believe that publication would greatly reduce developed countries' willingness to participate fully in the exercise. The compilation of such an inventory would also make it easier to identify major gaps, that is, types of equipment or expertise which are not readily available from developed countries. Findings of that kind would point to a need for further search and research.

The research budgets of the agencies studied are heavily concentrated on a small number of institutions in developed countries whose main focus is on the development and promotion of technologies which they consider appropriate to developing countries. However, a wide range of appropriate technologies may already be in use in developing countries, but not widely known. What seems to be needed, therefore, is a set of inventories on a national or regional basis. These can clearly be most efficiently undertaken by local institutions, and indeed a significant amount of activity of this kind is already taking place. To encourage such efforts, there may be a case for development agencies to switch part of their research financing to Third World institutions. It goes without saying that some diversification of subjects would also be desirable, since the present concentration on energy technology seems excessive.

Institutions concerned with information on appropriate technology have been enthusiastic about building up data banks, which tend to be underutilised, or utilised primarily by other information-gathering institutions rather than by project sponsors. So there may be a case for more operational activities aimed at assisting project sponsors in identifying what is available at the moment when they are ready to purchase. Such activities are conspicuously missing from the services currently provided by official development agencies. Some private institutions do offer such services, but only within the limits of their own particular interest. Admittedly, we do not pretend that the case for a wider-ranging service of this kind has been sufficiently made by the present study. The information problem is clearly one that merits further investigation.

NOTES

1. This study is a shortened and somewhat updated version of a paper written in 1979 under contract with the ILO. The author subsequently joined the Secretariat of the OECD. The updating that has been done for the present version concerns only matters of fact. (The examples of problem cases given in the first few pages were supplied by Fred Fluitman, of the ILO Secretariat.) Views and conclusions remain those expressed by the author in his personal capacity when under contract to the ILO.
2. Unless otherwise specified, we use the term 'aid' to refer to official development assistance (ODA) and other official flows controlled by bilateral or multilateral aid agencies. ODA is defined by the OECD as the sum of grants and loans for developmental purposes provided by the official sector at concessional terms (i.e., with a grant element of 25 per cent or more). Other official flows include non-concessional lending, notably World Bank loans, and official export credits.
3. See Westwood (1966) and Walters (1970). Ways in which developmental, commercial and political considerations may in fact coincide are

discussed in White (1974), especially chapters I, II and VIII. Additional references on donors' motivations are Little and Clifford (1965) and Mikesell (1968).

4. See OECD, DAC Report (1969), p. 214.
5. See Eckaus (1977), p. 79.
6. See ILO (1980), pp. 7–8. That aid discourages indigenous effort by providing an easy option is one of the standard criticisms of aid. For some vigorous examples, see Byres (1972).
7. See Coulsen (1978), p. 33. For aid projects in Tanzania, see also Green (1978).
8. See Randeni (1978), pp. 107–60 and 185–204.
9. See Radetzki (1977) and Beckman (1978).
10. See Thomas (1975), p. 55.
11. For a more thorough review of the rules and practices of the DAC member states' aid agencies, see OECD (1981).
12. See 'Die Entwicklungspolitische Konzeption der Bundesrepublik Deutschland' (Fassung 1975), in Bundesministerium für wirtschaftliche Zusammenarbeit (BMZ), *Entwicklungspolitik: Dritte Bericht zur Entwicklungspolitik der Bundesregierung*. Canada, Canadian International Development Agency, *Strategy for International Development Co-operation, 1975–80*.
13. See Sweden: 'Sweden's Policy for Co-operation with Developing Countries: A Summary of the Report of the Commission for the Review of Sweden's International Development Co-operation'. – Netherlands: 'Bilateral Development Co-operation: Concerning the Quality of Netherlands Aid', a note presented to the Parliament in September 1976 by the Netherlands Minister for Development Co-operation, Mr J.P. Pronk.
14. The United Kingdom, Ministry of Overseas Development: 'Overseas Development: The Changing Emphasis in British Aid Policies: More Help for the Poorest'. The United States: a series of legislative documents and reports under the general heading of 'New Directions' in development aid.
15. The World Bank: 'Appropriate Technologies in World Bank Assisted Projects Approved by the Board in 1977', a mimeo-graphed document.
16. IDB: 'Progress Report on the Application of Intermediate or Light-Capital Technologies in the Inter-American Development Bank', January 1978.
17. Quoted in OECD: DAC Report (1978).
18. Ibid.
19. See Singer (1965).
20. See Mason and Asher (1973), p. 235.
21. See Little and Mirrlees (1968).
22. See Squire and van der Tak (1975).
23. See Theierl: 'Technologies for Developing Countries', Deutsche Stiftung für internationale Entwicklung, Interregional Seminar on Selected Aspects of Technology Transfer, July 1977.
24. See World Bank: 'Development Finance Companies: Sector Policy Paper', April 1976.
25. During the present study, the writer served as a consultant to the

African, Caribbean and Pacific Secretariat in the negotiation of the second Lomé Convention. The assessments given here are largely based on work undertaken during that assignment.

26. Cf. Tendler (1975), chapter 7. See also Wildavsky (1964).
27. One of the earliest statements emphasising the seriousness of the problems caused by tying was Clifford (1966). See also Tendler (1975), Mahbub ul Haq (1967), Lall (1967), Bhagwati (1970), Miyamoto (1974), Wall (1973) and White (1974).
28. See OECD: DAC Report (1978), table B.4.
29. See OECD: DAC Report (1976), p. 159.
30. See Hills (1978).
31. There appears to have been no major progress in the evaluation methodology of technical corporation since the late 1960s (cf. OECD: *The evaluation of technical assistance*, OECD Technical Assistance Evaluation Studies, 1969.) At an OECD Development Centre seminar on the use of consultants in the mid-1970s, the focus was still on conditions for ensuring that consultants' advice will be accepted. The principal requirement was considered to be to allow consultants to write their own terms of reference!

REFERENCES

Beckman, B. (1978) *Aid and Foreign Investment: The Swedish Case, Part I*, Working Group for the Study of Development Strategies (AKUT), Uppsala University, December.

Bhagwati, J. (1970) 'The Tying of Aid', in J. Bhagwati and R. Eckaus, *Foreign Aid* (Harmondsworth: Penguin).

Burch, D. (1979) 'Overseas Aid and the Transfer of Technology: A Study of Agricultural Mechanisation in Sri Lanka', unpublished Ph.D. thesis, University of Sussex, 1979, abstracted in: *Development Research Digest*, IDS (Sussex), No. 3, 1980.

Byres, T.J. (ed.) (1972) *Foreign Resources and Economic Development: A Symposium on the Report of the Pearson Commission* (London: Frank Cass).

Canada, *Canada and Development Cooperation*, CIDA Annual Review, various years.

Clifford, J. (1966) 'The Tying of Aid and the Problem of Local Costs', *Journal of Development Studies*, January.

Coulsen, A. (1977) *The Automated Bread Factory*, University of Dar es Salaam, Department of Economics, April (Mimeo).

Coulsen, A. (1978) 'The Silo Project', *IDS (Sussex) Bulletin*, August.

Eckaus, R.S. (1977) *Appropriate Technologies for Developing Countries* (Washington DC: National Academy of Sciences).

Green, R.H. (1978) *The Automated Bakery: A Study of Decision Taking Goals, Processes and Problems in Tanzania*, the Institute of Development Studies at the University of Sussex, Working Paper No. 141, October.

Hills, John (1978) 'The European Development Fund: Proposals for the

Renegotiation', in *Renegotiation of the Lomé Convention: A Collection of Papers* (Catholic Institute for International Relations and Trocaire).

Jéquier, Nicolas and Blanc, Gérard (1983) *The World of Appropriate Technology* (Paris: OECD Development Centre).

ILO (1980) *International Labour Conference, 66th Session, Provisional Record*, No. 27 (Geneva).

Kreditanstalt für Wiederaufbau, *29 Jahresbericht, Geschaftsjahr 1977*.

Kreditanstalt für Wiederaufbau, *28th Annual Report covering the year 1976*.

Lal, D. (1967) *Cost of Aid Tying – A Case Study of India's Chemical Industry* (New York: UNCTAD).

Little, I.M.D. and Clifford, J.M. (1965) *International Aid* (London: George Allen and Unwin).

Little, I.M.D. and Mirrlees, J.A. (1968) *Manual of Industrial Project Analysis in Developing Countries* (Paris: OECD Development Centre).

Mahbub ul Haq (1967) 'Tied Credits: A Quantitative Analysis', in J. Adler (ed.) *Capital Movements and Economic Development* (London: Macmillan).

Mason, E. and Asher, R. (1973) *The World Bank since Bretton Woods* (Washington DC: Brookings Institution).

Mikesell, R.F. (1968) *The Economics of Foreign Aid* (London: Weidenfeld and Nicolson).

Miyamoto, I. (1974) 'The Real Value of Tied Aid: The Case of Indonesia in 1967–69', *Economic Development and Cultural Change*, April.

Netherlands (1976) *Implementation and Vindication of Policy in 1975*.

OECD, *Development Cooperation: Efforts and Policies of the Members of the Development Assistance Committee* (Paris) (to be quoted as DAC Report), various years.

OECD (1975) *Aid Evaluation: The Experience of Members of the Development Assistance Committee and of International Organisations* (Paris).

OECD (1981) *Compendium of Aid Procedures, A Review of Current Practices of Members of the Development Assistance Committee* (Paris).

Radetzki, M. (1977) 'Bai Bang: A Misguided Venture in Development Assistance', *Svenska Dagbladet*, 24 December.

Randeni, A.C. (1978) *The Influence of Aid on the Choice of Technologies in Industry*, chapter V of an unpublished Ph.D. thesis, University of Sussex.

Singer, H.W. (1965) 'External Aid: For Plans or Projects?', *Economic Journal*, September.

Squire, L. and Tak, H.G. van der (1975) *Economic Analysis of Projects* (Baltimore: Johns Hopkins Press).

Tendler, Judith (1975) *Inside Foreign Aid* (Baltimore: Johns Hopkins Press).

Thomas, J.W., in Timmer *et al.* (1975) below.

Timmer, C.P., Thomas, J.W., Wells Jr., L.T., and Morawetz, D. (1975) *The Choice of Technology in Developing Countries: Some Cautionary Tales* (Cambridge, Mass.: Harvard University Press).

Wall, D. (1973) *The Charity of Nations: The Political Economy of Foreign Aid* (London, Macmillan).

Walters, Robert S. (1970) *American and Soviet Aid: A Comparative Analysis* (Pittsburgh, Pa: University of Pittsburgh Press).

Westwood, Andrew F. (1966) *Foreign Aid in a Foreign Policy Framework* (Washington DC: Brookings Institution).

White, John (1974) *The Politics of Foreign Aid* (London: Bodley Head).

Wildavsky, Aaron (1964) *The Politics of the Budgetary Process* (Boston, Mass: Little, Brown and Co.).

World Bank (1974) *Uses of Consultants by the World Bank and its Borrowers*, April.

World Bank (1977) *Guidelines for Procurement under World Bank Loans and IDA Credits*, March.

8 The Patent System and Indigenous Technology Development in the Third World

SUSUMU WATANABE[1]

PROBLEM SETTING

The patent system in the broader sense including the utility model system[2] is probably the most conventional and also one of the most powerful policy instruments that have been used for the development and diffusion of new technologies. Together with Venice, England is one of the two lands of origin of the patent system. Here, a patent system was introduced under Edward III in the early fourteenth century for the above purpose. In the second half of the sixteenth century under the reign of Queen Elizabeth I, William Cecil, Secretary of State and later Lord Treasurer, 'was exceedingly anxious to develop English industry of every kind, so that the country might not only become economically independent, and be able to dispense with some of its imports, but might also have valuable commodities to export to foreign markets. The best hope of bringing about a considerable improvement at small cost, lay in the granting of patents to men who had enterprise enough to plant a new art, or introduce a new manufacture'.[3] A remarkable example in modern times is related to Japanese industries, where a recent survey discovered that the patent system was considered to be a primary incentive to industrial invention by firms.[4]

In the context of development economics, however, the existing works on the patent system are not intended to explore the possibility

217

of enlarging the scope of its contribution, but, rather, to restrict its use and reduce the cost of importing foreign technologies by trimming the monopoly power of patentees in industrialised countries. There are a number of reasons which are given to justify such an attitude. First, the Third World countries account for only a marginal proportion of patents granted in the world. Second, the patents with significant industrial application in these countries are held almost entirely by foreign corporations. Third, the situation has been worsening at least in some countries. Monopoly power has been abused by patentees in industrialised countries, while the stimulus to inventive activity has been marginal in developing countries.[5]

In the light of the historical evidence in industrialised nations, however, the validity of this negative attitude is not convincing. Isn't the frustrating experience of developing nations attributable to the inappropriate operation of the patent system and to the lack of some complementary policy or administrative machinery, rather than to any inherent weakness of the system? This question seems all the more relevant, because the critics of the patent system have not come up with any promising proposal of an alternative instrument for the encouragement of large-scale participation of the indigenous population in inventive activity (i.e., generation and adaptation of technologies).

Technological self-reliance of developing nations cannot be attained merely through free or low-cost access to technologies developed elsewhere (which may, in fact, delay the achievement of the objective as will be discussed later), but it presupposes formation of their own inventive capacity. It is then strange that the discovery of appallingly low percentages of indigenous inventions in developing countries has induced criticisms of the patent system and not a serious analysis of the causes of its failure. The present inquiry addresses itself to this very issue.

Before we start our discussion on the main subject, it is necessary to clarify what kinds of technology we are talking about. The critics are usually concerned with research-based, relatively large-scale modern technologies.[6] For the supply of such technologies, the developing nations rely mostly on imports, although countries like Argentina, Brazil, China, India, the Republic of Korea and Mexico have built up some indigenous technology base especially in areas related to defence (e.g., nuclear fusion, aerospace, electronics and certain fields of engineering). In the long run, progress made in such areas can contribute to the development of other industries.[7] For the purpose of reducing poverty, however, developing nations need to improve the efficiency of medium and small enterprises manufacturing an endless variety of consumption and capital goods for domestic and export markets. Our inquiry is mainly concerned with technologies which are useful for this purpose.

Effort at the development of the second type of technologies appears to have been inadequate even in the technologically more advanced of the developing countries, except the Republic of Korea. The situation may be illustrated with reference to India, which is now claimed to be the leading exporter of technologies among the Third World nations.[8] On the one hand, the government can report: 'In the area of application of science, success has been achieved in several fields in agriculture and in specific mission-oriented specialised agencies for atomic energy and space.' On the other, it is concerned that 'In large areas of economic activities, relatively obsolete cost-ineffective technology continues to be applied, the pace of scientific and technological innovation remains unimpressive and the adoption of the available scientific and technological knowledge is tardy.'[9] One result is a delay in industrialisation and persistent poverty, as is clear from Table A.1 in the Appendix to this chapter.

Even those authors who stress the importance of indigenous supply of technologies usually talk about formal R&D programmes and rarely touch on the non-institutional inventive capacity.[10] However, experience in the last few decades indicates that a government can do little *directly* to reduce mass unemployment and poverty, especially in market or mixed developing economies. Its technological, financial and administrative capacity is far too small for the task. This seems particularly true with regard to the type of industrial technologies under consideration: 'the incredible heterogeneity of manufacturing industries renders direct government assistance difficult'.[11] Then, the low level of inventive activity by indigenous populations should be a cause of even greater concern than the poor performance of formal R&D institutes, which has been a frequent subject of study.

The rest of this chapter will first examine the level of inventive activity in different countries using patent statistics. It will be shown that international differences in this respect can be explained broadly by differences in the level of industrial production and in the rate of adult literacy, as is predicted from theories of invention. The third section briefly reviews the structure and problems of R&D systems in the Third World. The fourth section deals with factors that seem to influence people's enthusiasm and capacity for inventive activity. In the fifth section, pre-war Japan's experience will be briefly examined, by way of exploring promising areas of government action to encourage inventive activity in Third World countries. In the final section, implications of our findings for policy-makers and researchers will be considered.

THE LEVEL OF INVENTIVE ACTIVITY IN THE THIRD WORLD

Inadequacy of patent statistics as an indicator of *innovative* activity is well known.[12] However, it has been widely accepted as almost the only practical indicator of *inventive* activity, despite its shortcomings.[13] Inter-temporal and international variations, for example, in the product groups covered by the patent system, forbid use of readily available patent statistics for a very precise comparative analysis. Still, one may be able to discuss the relative level of inventive activity among different nations, where observed gaps are large enough.

The number of patents granted can be influenced considerably by the administrative capacity of the patent office. Patents are granted without examination in some countries (France and Switzerland among the industrialised countries in Table 8.1), while patent applications are subject to screening in other countries. Moreover, a given number of patents or patent applications naturally have different significance depending on the size of population. In the present study, therefore, the level of inventive activity or inventiveness of a nation is measured by the number of inhabitants per patent application: the smaller the number, the greater the inventiveness of a nation. In Table 8.1, it is given together with basic patent statistics for 1979. Some industrialised countries are included for the purpose of comparison.

This table does not include the centrally planned economies and fifteen developing countries which are listed in the original WIPO table: Bahrain, Burundi, Ghana, Honduras, Kenya, Mali, Mauritius, Peru, Rwanda, Samoa, Seychelles, Sierra Leone, Singapore, Uganda and Tanzania. In 1979, there was no application filed by the residents (not necessarily citizens) in these countries. Some of them (e.g., Singapore and Kenya) traditionally file patent applications not locally but at the British patent office. Algeria issues patents for foreigners and inventors' certificates for her citizens, as is the practice in Eastern European countries. Five applications for inventors' certificates were

TABLE 8.1 *Patent applications and grants during 1979*

	Patents granted Total (1)	Residents (2)	Patent applications filed Total (3)	Residents (4)	Inbabitants per application (5)
Argentina	3 375	1 244	4 482	1 314	20 786
Bahamas	37	1	37	1	n.a.
Bangladesh	103	20	131	31	2 828 000
Bolivia	127	20	134	15	361 866
Brazil	1 583	175	8 602	1 958	59 519
Colombia	844	36	420	45	580 488
Costa Rica	20	6	107	30	72 066
Cyprus	52	2	52	2	n.a.
Ecuador	110	7	170	23	348 739
Egypt	376	6	784	61	637 180
El Salvador	67	6	143	16	275 625
Hong Kong	893	26	848	17	292 058
India	2 182	594	2 910	1 053	626 391
Indonesia			477	12	11 905 833
Iran (Islamic Rep. of)	1 061	11	820	83	445 433
Iraq	82	9	220	37	341 378
Jamaica	63	2	73	6	359 833
Korea (Rep. of)	1 419	258	4 722	1 034	36 570
Malawi	37		37	2	2 908 500
Malta	12	2	19	2	n.a.
Mexico	2 026	236	4 485	692	94 666
Morocco	372	31	391	29	673 724
OAPI*	571	26	284	6	n.a.
Pakistan	446	8	404	30	2 656 833
Philippines	857	82	1 471	144	324 638
Sri Lanka			227	53	274 377
Thailand			22	7	6 496 428
Tunisia	242	3	261	26	238 230
Turkey	458	34	558	73	605 986
Uruguay	108	15	213	43	67 534
Venezuela	660	39	2 115	192	75 276
Zaïre	96	11	98	11	2 500 818
Zambia	53	—	95	1	5 580 000
Zimbabwe	183	12	256	55	129 927
France	24 618	6 846	32 174	11 303	4 722
Germany (Fed. Rep.)	22 534	10 895	55 184	30 879	1 980
Japan	44 104	34 863	174 569	150 623	768
Switzerland	6 614	1 638	11 540	4 441	1 454
United Kingdom	20 800	4 182	44 666	19 468	2 873
United States	48 853	30 605	100 494	60 535	3 694

SOURCES   Columns (1), (2), (3) and (4); WIPO (1981).
Column (5): Population figures in *1981 World Bank Atlas* were divided by the figures in (4).

NOTE   *12 member states of the African Intellectual Property Organisation: Benin, United Republic of Cameroon, Central African Republic, Chad, Congo, Ivory Coast, Mauritania, Niger and Upper Volta.

filed in 1979. Mexico issues both patents and inventors' certificates for residents and non-residents. There were thirty-four applications for the latter by residents and 655 by non-residents in the same year.

The number of non-residents' applications for patents exceeds that of residents' almost universally. The only exceptions are the Federal Republic of Germany, Japan, and the United States. However, the gap between the two is proportionally much wider in developing countries than in industrialised countries. What is more, differences in the number of inhabitants per application between the two groups of countries and among the developing countries are too large to be ignored.[14]

Only ten countries or so have a utility model system. Data from eight countries (Brazil, Federal Republic of Germany, Japan, Republic of Korea, the Philippines, Poland, Portugal and Spain) are reported most regularly in the WIPO publication. Italy and Liechtenstein appear from time to time. Residents are predominant regarding both applications and grants in every country. In 1979, Brazil, the Republic of Korea and the Philippines filed 1770, 7957 and 786 applications, respectively. In each case, residents' share was 1690, 7215 and 762. This contrasts sharply with the case of the patent system, and suggests a greater impact of the utility model system on indigenous inventions. It is also interesting that the Republic of Korea and Brazil which are often considered to be almost completely dependent on imported technology appear to have been even more enthusiastic about encouragement of indigenous inventive effort than those countries which are usually considered to be more independent technologically (e.g., India).

How can the international variation in the level of inventive activity be explained? Theories of invention established in industrialised countries may provide some clue to answer the question.

The first relevant theory is that, basically, invention is an economic activity motivated by economic gains. Once in a while, an important invention is made as a result of the insight and intuition of a genius. Some people show enormous enthusiasm for inventive activity, simply because they find great pleasure in it regardless of economic returns. But one may safely treat them as exceptional cases. Second, an invention can be either autonomous or induced. The first type of inventions are the products of the inventor's independent mind with little relationship to industrial activities in which they have been involved. The second type of inventions appear in response to recognised needs for improvement of production methods or products which are currently in use or in production. Independent individuals are more likely to be associated with an autonomous invention, while corporate inventors (e.g., firms' R&D units) will naturally be concerned with technological problems related to the firms' lines of production. Independent inventors account for a significant percentage of epoch-making inventions. In contrast, large industrial research laboratories are a relatively minor source of such inventions but a main source of improvement (or induced) inventions. In terms of the number of inventions, there is a marked trend towards increasing relative importance of corporate inventions, after the Second World War. In the United States, 80 per cent of patents were issued to individuals in 1900 as compared with 30 per cent around 1970. The situation in the United Kingdom is much the same.[15] In Japan the corporations accounted for 88.2 per cent of residents' patent applications and 79.9 per cent of residents' utility model applications in 1977, while individuals' share was 9.8 per cent and 19.7 per cent respectively.[16] This implies a rising trend of problem-solving induced inventions.

The numbers of induced inventions and related patent applications tend to grow with aggregate industrial production. This proposition has been verified by Schmookler's empirical study in the United States.[17] He has discovered a close relationship between the number of patent applications and the level of industrial production or investment: the former fluctuates following ups and downs in the latter with some time lag. The demand–pull aspect of inventive activity has also been confirmed by a study in Japan where patent applications followed fluctuations in industrial production with a few years' time lag over six decades starting in the mid-1880s.[18]

Some authors argue that Schmookler somewhat overemphasised the demand factor. For example, Scherer maintains, on the basis of empirical studies in the United States, that differences in technological opportunity – for example, differences in technical investment possibilities typically opened up by the broad advance of knowledge – are a major factor responsible for inter-industry differences in inventions.[19] Referring to some broad historical evidence, Rosenberg[20] stresses the importance of technological possibilities as a major determinant of the level of inventive activity in different areas of invention: for example, progress in medicine was very small until great breakthroughs were made in bacteriology by Pasteur and Lister, despite the persistence of the need for such progress over centuries.

Obviously, both technological possibility *and* the demand factor influence the probability of success and cost of inventive activity within any particular period. Moreover, the level of technical knowledge or the technological possibility of invention is at least partially dependent on the level of production in the recent past through a 'learning effect'. Also, areas of inventive activity will increase with the diversity of industrial production since 'men invent in the fields they are familiar with, and they are most familiar with the goods they produce and consume'.[21] Growth of output may also necessitate technological progress aimed at higher productivity to overcome supply constraints. At the same time, R&D funds tend to grow as the level of industrial production rises.

From these empirical studies in industrialised countries, it may be hypothesised that international differences in the level of inventive activity are attributable to differences in the level of income or industrial production as well as in the level of education or literacy. To test such hypotheses, we examined the correlation between the number of inhabitants per patent application as a reciprocal of the level of inventive activity and five other variables: the secondary school enrolment rate, the higher level education enrolment rate with respect to population aged 20–24, the adult literacy rate, GNP per capita and gross domestic manufacturing product (GDMP) per capita. Our reference year is around 1979, as indicated in Table A.1 of basic data annexed to this chapter. This choice was made mainly on the basis of the availability of population data. For the non-patent statistics, we relied on World Bank data. This obliged us to exclude a few countries with fewer than one million inhabitants from our analysis, although the WIPO publication gives their patent statistics. The correlation coefficients were calculated using non-transformed variables with respect to education and literacy, and natural logarithms of the three remaining variables (i.e., inhabitants per patent application, and GDMP and GNP per capita). In all cases, results were significant at the 0.01 per cent level, and the sign of the coefficient was negative, indicating a positive correlation between the level of inventive activity and the selected variables. The squares of the correlation coefficients are given in Table 8.2.

Clearly, the correlation between the level of inventive activity and that of GDMP per capita is positive and strong (Figure 8.1). The case of GNP per capita is similar. These findings confirm that the relationships established between patent applications and industrial production in industrialised countries holds good internationally, as well.

TABLE 8.2 *Correlation between the number of inhabitants per patent application and selected variables*

	Variables		r^2
(1)	GDMP per capita	(53)	0.82
(2)	GNP per capita	(60)	0.80
(3)	Adult literacy rate	(52)	0.67
(4)	Secondary school enrolment rate	(60)	0.75
(5)	Higher education enrolment rate of population aged 20–24	(58)	0.58

NOTE The figures in parentheses indicate the number of countries included in the calculation.

The correlation between the inventiveness of a nation and the three selected variables on education and literacy was found to be somewhat weaker, but still fairly strong. The correlation is strongest with secondary school education and weakest with higher education. This suggests that lower level education has greater relevance to the development of a nation's inventive capacity than university and college level education.

Needless to say, the level of education and literacy is not independent of GNP or GDMP per capita. Although the relationship is not quite unidirectional, by and large, a rising income level facilitates spread of primary and secondary education in the Third World. Accordingly, expanding national and industrial production may have a much greater positive impact on the level of inventive activity in developing countries, because it not only creates a demand–pull effect on such activity but also enlarges the technical capacity to invent, to a much greater extent than in industrialised countries.

Schmookler found a high correlation between R&D expenditure and patent applications in his 1966 study cited above. Omission of this subject from our analysis is primarily due to the paucity of relevant data. UNESCO publishes data on R&D expenditure in some developing countries, but their coverage varies from country to country: for example, R&D expenditure in the private sector is often left out.

Internationally, however, the impact of R&D expenditure will vary because of the difference in the pattern of inventive activity. The amount of such expenditure must have a decisive influence on basic research and major inventions, but more limited relevance to applied research and improvement inventions. For example, Japan's total

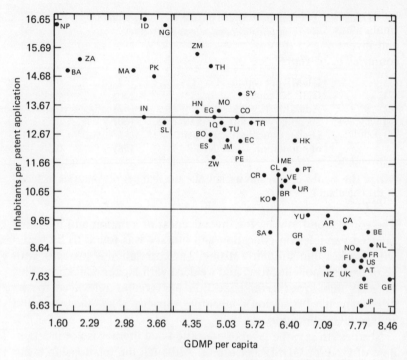

SOURCE: Based on the data in Table A.1 in the Appendix.

FIGURE 8.1   *Correlation between the number of inhabitants per patent application and GDMP per capita in 1979 (in logarithms of the variables)*

R&D expenditure was about one-third of similar expenditure in the United States around 1977–8, excluding R&D related to defence.[22] If this factor is taken into account, the gap between the two countries was far greater. However, the total number of patent applications is 74 per cent greater in Japan as shown in Table 8.1. (The smaller number of patents granted in Japan is attributable, at least partly, to the limitation in the Japanese Patent Office's capacity to process applications.) The low cost of Japanese inventions reflects the well-known bias of the Japanese R&D programme towards technology adaptation and improvement.

In most developing countries, modern industries characterised by large optimum scale of production and high technology represent islands in the ocean of traditional and small industries. Often owned by foreign firms or the government, they depend on imported technology. The numbers of people working in these industries are still small. Thus, one may broadly expect typical indigenous technologies and inventions in these countries as well as those in industrialised countries to be represented as in the following scheme:

	Developing countries	Industrialised countries
Indigenous technology	(a) conventional (b) low technology (c) small scale	(a) modern (b) high technology (c) large scale
Invention	(a) experience-based (b) personal knowledge-based	(a) science-based (b) system-based
Inventor	(a) individuals or (b) small firms	(a) R&D institutes or (b) large firms
Imitation	(a) easy	(a) difficult

For the type of technologies which one would expect from the indigenous population of a developing country, R&D expenditure seems to have much more limited relevance than in industrialised countries, although its importance will naturally increase as the nation's industrialisation advances.

R&D ACTIVITY IN THIRD WORLD COUNTRIES: ITS STRUCTURE AND PROBLEMS

In most developing countries, formal R&D activity is mainly confined to the government and foreign-controlled sectors. The government's R&D expenditure is intended partly to meet its own needs (e.g., defence) and partly to make good the lack of effort by the private sector. Foreign firms do try to adapt technologies brought from their home countries and to look for local substitutes for imported materials, when this is both technically and economically feasible under the local conditions, and especially when they are pressed by the host government to do so.[23]

The heavy involvement of the government in the national R&D programme is not rare in industrialised countries. The government accounts for 50 per cent or more of the national R&D expenditure in the United States, the United Kingdom and France, as noted earlier. Significant proportions of the official R&D funds are related to the defence programme. In these countries, however, the relationship between R&D institutes and industry is much closer than in most of the developing countries. Moreover, the distinction between defence and civil industries is not clear, technologies developed for the former being put to use in the latter fairly quickly. This is so, largely because the mainstay of an industrialised economy is research-based modern industries. In a developing economy with a considerably different industry-mix, the concentration of R&D activity in the government sector, which is even greater than in the industrialised countries, gives rise to peculiar problems. This may be illustrated with reference to India.

R&D expenditure in India grew from 229.3 million rupees in 1958–9 to 1075.6 million rupees in 1968–9 and to 4022.5 million rupees in 1976–7, roughly a five-fold increase in real terms. Its ratio to GNP rose from 0.23 per cent to 0.66 per cent over the period. The private sector's contribution increased considerably: 1.5 million rupees (0.7 per cent of the total) in 1958–9, 98.5 million rupees (11.5 per cent) in 1968–9, and 495.0 million rupees (15.4 per cent) in 1976–7.[24]

The government supports some forty laboratories which are intended to develop and supply indigenous technologies to local industries. They are under the control of the Council of Scientific and Industrial Research (CSIR). The National Research Development Corporation (NRDC) is responsible for marketing of technologies generated by these and other government-run laboratories, sometimes providing loans or equity capital to enterprises which buy a technology. A number of technologies developed by these laboratories have been sold abroad (Suri transmission, improved aluminium based alloy, activated carbon, and high draught kiln) and some others have attracted the interest of foreign firms.[25] Some institutes, such as the Central Leather Research Institute (Madras), have been more successful than others, often because of their closer tie with specific industries of the region. However, the overall performance of these laboratories has been considered disappointing, and they have been criticised on many occasions. They developed slightly over 2100 technologies between 1954 and March 1980, and some 450 of them had been put into operation by the end of 1980.[26] But the government has incurred a heavy financial burden: in 1974–5, for example, the laboratories spent 194 million rupees while the NRDC earned only 4.3 million rupees from the sale of know-how and technical services.[27] Consequently, in December 1979, the governing body of the CSIR urged the need for investigation into the causes of the low rate of commercial application of technologies generated by these national laboratories.[28]

As a matter of fact, a study published in 1971[29] had already noted that 'much industrial work done by CSIR laboratories either did not reach a marketable form, or when considered to have reached such form, did not find a buyer, or was sold below cost'. It explained the situation as follows: (i) the know-how from the laboratories was obsolete, probably because the time element was neglected by them; (ii) laboratories made prototypes regardless of cost and import content; (iii) laboratories, having no manufacturing experience, could not give industrialists as much help as foreign technology suppliers to solve problems of blowing up of processes, start-up, and adaptation.[30]

The very basic cause of the problem seems to be this: 'the CSIR laboratories work . . . on the basis of an almost total isolation from day-to-day problems in productive enterprises'.[31] In principle, their research projects are approved on the basis of their practical relevance. However,

the practical relevance quite often remains in theory only, for science in India is not strictly related to production. Prospective users of industrial research either do not know of the Indian know-how or deliberately ignore it in favour of foreign technology for reasons of expediency, risk coverage, prestige and other economic and cultural reasons. The Indian scientist has not been politicised and ideologically motivated to work toward national goals. He considers himself a scientist first and foremost, above and beyond politics or even society.[32]

Little market information goes into the choice of research projects. Consequently, Indian firms look to foreign companies for supply of technologies, although, to eliminate competition with imported technologies, a CSIR representative has been on the Foreign Investment Board, which approves technology imports, since the late 1960s.[33]

These problems are far from unique to India. For example, a Caribbean and Latin American symposium came to the conclusion that a major problem in many of the countries in the region is the preoccupation with science and technology as ends in themselves, rather than as tools of development. In its proceedings, Wionczek writes as follows:

Mexican experience strongly suggests that the supply of internally produced scientific knowledge and technical know-how does not create an automatic demand for them, because historically the demand has been directed to the outside. Consequently, the advancement of science and technology in an underdeveloped country depends more upon the effort to establish links between the R&D system and the education system and the economy, than upon the simple increase in human and financial resources allocated for R&D.[34]

Most of the problems of the official R&D programmes will be avoided in similar programmes of private sector enterprises, simply because they cannot afford to undertake projects unrelated to their business and also because they know exactly what is needed in their production lines.

In the case of India, the government became more restrictive regarding the import of technology after 1966. More recently its income tax law has provided that one-third more than the actual R&D expenditure can be allowed for in computing taxable profits. Consequently, companies have been encouraged to increase their R&D expenditure. However, many Indians believe that a considerable proportion of the 'R&D expenditure' declared has not really been used for R&D activities, but for mere quality control facilities and 'R&D buildings' which play the role of warehouses. Nevertheless, these incentives provided by the government together with the increased R&D expenditure in the public sector are perhaps responsible for the rising trends of patent applications by, and grants to, Indian residents (Table 8.3).

It must be remembered, however, that, as in many other developing countries, R&D expenditure in the private sector in India has been attributable largely to foreign-based companies. For example, in the pharmaceutical industry (which is excluded from the patent system in India and other developing countries),[35] the largest spender among the private sector industries in India, subsidiaries of multinational enterprises accounted for 82.6 per cent of the total R&D expenditure in 1972/3–1974/5. Their research was aimed at the adaptation of technologies obtained from the principal firms.[36] Since foreign-controlled enterprises are legally treated as national corporations in many countries including India, the patents granted to those companies are regarded as national patents. It follows that the rate of

TABLE 8.3 *The average annual number of patent applications and acquisitions in India (1945–1979)*

| | Applications | | | | Acquisitions | | |
	Indians	Other Residents	Foreigners	TOTAL	Indians	Foreigners	TOTAL
1945–9	275	73	1 775	2 123	n.a.	n.a.	n.a.
1950–4	411	48	1 733	2 192	n.a.	n.a.	n.a.
1955–9	522	62	2 775	3 359	278	2 063	2 341
1960–4	737	81	4 579	5 397	321	3 073	3 394
1965–9	1 006	106	4 374	5 486	454	3 694	4 148
1970–4	1 130	149	2 748	4 024	514	2 241	2 755
1975–9	1 154	30	1 813	2 997	573	1 752	2 325

NOTE Since 1973 the relevant data have been published with respect to each fiscal year which starts on 1 April. To be consistent with the earlier data, adjustment has been made by shifting one-fourth of a fiscal year figure to the next.
SOURCE *Annual Reports* of the Patent Office.

genuinely indigenous applications may be considerably lower than Table 8.1 and the data in the Appendix suggest. Another cause of concern is that there is a sign of declining participation of individuals in inventive activity in some countries. In India, the percentage of patent applications by enterprises and research institutes (including the government) grew from 41 per cent in 1967–70 to 63 per cent in 1976–9.[37] In the light of the data in Table 8.3 this implies a decline in the absolute number of applications by individuals.

To reduce the gap between R&D products and users' needs, two approaches are usually advocated: (i) to increase contacts between the R&D institutes and potential users of their products (i.e., new technology); and (ii) to encourage popular participation in (informal) R&D activity. Considering the vastness of the problem (surplus labour and poverty) in most developing countries and the modest scale of the modern industrial sector in these countries, one may safely argue that their success in development effort depends more crucially on achievements in the second field.

From this viewpoint, the above pattern of inventive effort is worrying. Even more serious is the fact that only three relatively advanced developing countries are equipped with a utility model system to encourage people's inventive activity.

CAUSES OF LIMITED PARTICIPATION IN INVENTIVE ACTIVITY

In India where the patent system was introduced as early as in 1856, only about 270 patents were obtained annually by her citizens a century later (Table 8.3). Out of 162 leading Indian enterprises listed in the *Indian Engineering Industries*, only three owned substantial numbers of patents, and nine others a single patent each. Trying to explain such 'failure' of the system, an ex-Controller of Patents and Designs[38] argued that the system was something imposed by the British to extend protection for British patent-holders to India and therefore did not reflect India's own needs. He further maintained that, 'In these circumstances, it is not surprising that the Indian patent system has, in spite of its operation for a hundred years, failed to become popular in this country, but – like the Top Hat and Morning Coat of the Foreigner – is still looked upon as a foreign system.' This explanation was perhaps sensible not only in India but also in many other countries, notably in Latin America where the patent system often dates back to the nineteenth century. It may also be true that long colonial rule reduced self-confidence among potential local inventors. Decades after their independence, however, it is not convincing to attribute the limited inventive activity by their indigenous populations mainly to colonialism. There may be more fundamental explanations within their economic and social system or in their government policy.

Our analysis in the second section suggests that the stagnation of industrial production may be a major cause, as well as effect, of the lack of people's enthusiasm for invention. Another important explanation may be the lack of industrial competition. Producers would not get interested in technological innovation unless they face a need to do so. Often they are forced to look for new technologies to reduce the cost of production and improve the quality of products, when they are pressed hard by their competitors. In most developing countries, such stimulus is missing. In rural industries, petty producers often work for customers with whom they are socially tied. The latter are often ignorant about alternative sources of supply.[39] In modern industries, the size of the accessible market is usually too small to attract sufficient numbers of investors to create a healthy competitive climate. Not uncommonly, the slow growth of industries and lack of competition are partly a consequence of government interventions.[40] In India, all major companies have to apply for a licence to expand their production capacity beyond a certain limit. The purpose of this regulation is to control the concentration of economic power within an industry, but it may also discourage inventive effort. Multinational enterprises interviewed in a recent survey felt that it was difficult to invest heavily in R&D since they were not even sure that they could get a licence to use the resulting new technologies.[41]

As Schmookler has shown in his studies cited above, inventive activity is largely motivated by the expected profitability of inventions. Then, people's lack of enthusiasm for such activity may be basically due to their unawareness of income opportunity associated with inventions. This in turn may be attributable, at least partly, to the weakness of the patent system.

The patent system in developing countries is prone to three fundamental weaknesses. First, the system exists, but is not known to many people. During our recent conversation, a director of India's Patent Office said, 'In an industrialised country, even a school boy would know what a patent means, but here even industrialists do not know about it.' Whether the patent system can stimulate inventive activity or not depends partly on how far the patent office can disseminate information concerning useful inventions. In most developing countries the patent office is little more than a mere registry. The main problem is, of course, the budgetary constraint and consequent shortage of staff. Second, the protection expected under the patent system is often non-existent. A director of one of the largest company research institutes in India told the author that his company did not apply for patents to earn profits, because there was no effective means to protect patentees' rights in the country. They applied for patents for fear that someone else might apply for patent protection on the basis of their work and prevent them from using their own technologies. Shortage of staff may also be a major cause of frequent infringement of patent rights. Third, the patent system in developing countries often lacks complementary policy support. Like some good medicines, undesirable side-effects of the patent protection need to be controlled with other policy instruments, notably anti-monopoly legislation. Developing countries are rarely prepared adequately in this respect, as Vaitsos notes.[42] Policies concerning foreign investment and technology imports have also much to do with their unhappy experience with foreign patentees.

In connection with the last point, the following findings of the already cited NCAER study are quite suggestive:

Against India's 2,200 agreements, Japan entered into 6,400 technology import agreements between 1951 and 1964. Japan was particularly liberal towards short-period agreements, lasting no longer than a year, and generally involving the outright purchase of technology. By 1964 Japan had made nearly 3,400 such agreements as compared with 600 agreements of this kind in India. While only 3 per cent of India's agreements were for over 10 years, 20 per cent of Japan's were for 15 years. This liberal and selective policy probably enabled Japan to keep down reliance on the import of technology through foreign subsidiaries (p. 8).

At the time of the writing of the report, a royalty of up to 3 per cent was admissible and might go up to 5 per cent in exceptional cases. A royalty exceeding 5 per cent was not normally considered. The royalty paid per agreement was several times higher in Japan (p. 8). 'The inevitable result is that the more advanced techniques, which earn over 5 per cent in the rest of the world or which entail high royalties owing to India's small markets, are kept out of India. The Government's discouragement of export restrictions in collaboration agreements similarly tends towards the exclusion of relatively newer technology from the country' (pp. 77–8). In fact, 53 per cent of the agreements concluded by Japan up to 1961 contained export restrictions, the proportion for India up to 1967 did not exceed 7 per cent (p. 7).

These findings suggest that technological dependency on more advanced countries in the shorter run does not necessarily contradict the longer-term objective of technological self-reliance, but can provide a necessary base for the latter. Indeed, the authors of the above study note that Japan's heavy imports of technology in the 1950s were followed by a rapid rise in indigenous research and development (p. 79). A typical example is the iron and steel industry. This industry spent enormous amounts of money to import technologies in the 1950s and early 1960s, on the basis of which it has become a net exporter of technology in two decades. It earned 2.7 times more from technology exports than it spent on technology imports in 1981. On the national level, Japan is still a net importer, but the ratio of technology imports to technology exports (in terms of value) fell from 5:1 to 1.5:1 between 1970 and 1980. So long as new contracts are concerned, exports surpassed imports for the first time in 1972.[43] The Republic of Korea appears to be following a similar development strategy.

Use of imported technologies tends to necessitate 'problem-solving' inventions to adapt them to local conditions and needs. The greater the intensity of competition the users of the technology have to overcome, the greater their enthusiasm for such adaptive inventions will be (unless they get demoralised). It is also obvious that the lower the cost of imported technology, the smaller the incentive for importers to develop local substitutes. Just as the enormous cost of hiring foreign advisors and instructors stimulated Japan's desperate effort to replace them with local personnel in the 1870s and 1880s,[44] the heavy burden of her technology import bill lent a great impetus to her R&D programme. Thus, the mode of importation and application of foreign technologies seem to have a decisive influence on subsequent inventive activity.

While 'wants to invent' are influenced by all these factors, the 'capacity to invent', or the knowledge base essential for inventive activity, is developed largely through education and training. In this domain too problems of developing countries are well documented.

Their universities and colleges emphasise academic, instead of practical, training and research.[45] Thus, education at a university or college helps little in preparing a person for industrial activity. This may be part of the explanation for the low level of patent applications by Indian citizens despite the fact that the country now has 119 universities, five institutes of technology, 150 engineering colleges and about 100 medical colleges and 350 polytechnics, producing about 150 000 qualified scientific and technical people every year.[46] Similarly, Forsyth in this volume reports that Egyptian businessmen felt their compatriots' scientific effort 'too theoretical' and notes that investments are based heavily on imported technology, although two-thirds of all graduate degrees are in pure and applied sciences and there are over 260 research institutes, many attached to Ministries.

Second, the enrolment rate regarding higher education is disproportionately higher in developing countries, compared with the rate of literacy and the rate of enrolment in the secondary school (cf. Table A.1 in the Appendix). In the light of our findings in the second section, this is perhaps an even more serious problem.

INVENTIVE ACTIVITY IN PRE-WAR JAPAN

Japan's first national patent system was introduced in 1871. It was based on the model of the French patent law of 1844. As Korekiyo

Takahashi, father of the Japanese patent system, notes in his auto-biography, however, 'there was no adequate Japanese staff who could properly examine the applications. The Government was obliged to hire a large number of foreigners. The financial burden was enormous, and yet the technological standard of inventions was not high enough to justify such expenses'.[47] Consequently, this first attempt officially came to an end within a year. However, the Japanese enthusiasm for the patent system remained alive. Takahashi and others continued to study patent systems in Europe (especially France) and the United States, and a Monopoly Sales Patent Decree was issued in April 1885 to lay the cornerstone of the country's patent system. The new Decree resembled the French system in its format but in substance it was more like the American system.

When the first patent system was introduced in 1871, almost no modern economic and legal institutions had been set up, so much so that the feudal clans were replaced by the modern prefectural governments three months later! Between 1871 and 1885 the situation changed considerably. In the domain of industrial technology, the participation in the Vienna International Exhibition of 1873 and the government-sponsored study tours to various advanced countries which were undertaken on that occasion had created an enormous enlightening effect. The national industrial fairs of 1877 and 1881 further stimulated people's interest in technological progress for economic development.

Consequently, people's response to the new decree was swift. Before the end of 1885, 425 applications were accepted and 99 patents were granted. The number of applications rose from about 800 per year to over 1000 after 1889. By 1897 some 3000 patents were issued. Only 60 of them were granted to formally trained (university or secondary school level) engineers and technicians.[48] This is natural, because the scale of higher level education in Japan was still very modest as will be made clear shortly.

The Decree of 1885 was intended to encourage inventions by the Japanese and did not grant patents to foreigners. However, Japan had to agree to join the League of Paris Convention in the course of its negotiation for the revision of the 'unequal treaties'[49] imposed by the Western Powers in the mid-nineteenth century. Accordingly, a new Patent Law was enforced in 1899, two weeks before its entry to the League.

As Table 8.4 shows, the number of patent applications by Japanese citizens continued to grow under the new Law, and world-famous

TABLE 8.4 *Average annual number of patent and utility model applications and registrations in Japan* (1885–1979)

	Applications					Registrations		
	Japanese No.	%	Foreigners No.	%	TOTAL	Japanese No.	%	TOTAL
(A) Patents								
1885–9	911	100.0			911	161	100.0	161
1890–9	1 313	96.9	45	3.1	1 358	298	96.1	310
1900–9	3 212	86.5	502	13.5	3 714	963	70.8	1 360
1910–9	6 082	87.3	885	12.7	6 968	1 322	72.1	1 834
1920–9*	9 662	82.0	2 121	18.0	11 783	2 418	68.0	3 559
1930–9	13 996	86.3	2 226	13.7	16 222	3 813	77.4	4 929
1940–5	n.a.	n.a.	n.a.	n.a.	15 021	n.a.	n.a.	6 239
1946–9	10 561	97.7	250	2.3	10 811	2 290	98.7	2 321
1950–9	23 518	80.9	5 530	19.1	29 048	5 647	73.4	7 695
1960–9	55 384	73.4	20 059	26.6	75 443	14 843	66.1	22 462
1970–9	121 602	82.0	26 764	18.0	148 366	32 211	76.7	41 999
(B) Utility models								
1905–9	8 885	99.9	7	0.0	8 892	3 212	99.9	3 214
1910–9	15 044	99.9	17	0.1	15 061	3 508	99.9	3 514
1920–9*	25 769	99.6	88	0.4	25 857	7 679	99.5	7 721
1930–9	36 451	98.9	376	1.1	36 827	13 906	98.4	14 141
1940–5	n.a.	n.a.	n.a.	n.a.	22 927	n.a.	n.a.	12 824
1946–9	16 793	99.9	10	0.0	16 803	3 294	99.6	3 310
1950–9	48 748	98.8	575	1.2	49 323	14 542	98.3	14 794
1960–9	97 252	98.2	1 744	1.8	98 996	27 102	97.4	27 833
1970–9	161 001	98.9	1 737	1.1	162 738	41 294	98.2	42 084

*Annual average of nine years excluding 1922 for which no breakdown of the total figure is available.

SOURCE   Compiled using basic data given in Sangyô Kenkyûjo (Institute of Industrial Research): *Shin-gijutsu Kaihatsu ni okeru Jitsuyô Shinan Seido no Yakuwari* (The role of the utility model system in the development of new technologies) (Tokyo: 1981), pp. 242–5 based on annual reports of the Patent Agency, the Government of Japan.

discoveries and inventions began to appear, such as Jokichi Takamine's adrenalin (1901) and Taka-diastase (1909), Sakichi Toyota's automatic weaving loom (1907), and Umetaro Suzuki's oryzanin (1910). However, Japan was still at the final stage of its 'first industrial revolution' in textile industries. Accordingly, a large majority of the Japanese inventions registered with the Patent Bureau ('Patent

Office' after the Second World War) were partial improvements on imported machinery or traditional local equipment, or some other simple devices. Competition among users of such simpler technologies was intense, and improved ideas were copied quickly. Consequently, it was felt that those improvements and devices which were too minor to be protected by the patent law needed some protection from imitators. Hence the Utility Model Law of 1905 was introduced.

In preparing this law the Japanese studied Germany's utility model law of 1891. The Japanese law was similar to it in the following respects: (a) that the minor improvements and devices were to be protected not with a patent law or a design law but with a separate law, and (b) that the protection provided by this law was to be limited to a shorter period. In many other ways, though, it was original. First, while the German law was concerned with tools and equipment, the Japanese law protected 'useful new ideas concerning the form, structure or combination' of industrial products in general. Second, unlike the German law, it required an examination of each application before granting protection. Third, the degree of protection was much stronger under the Japanese law, since it provided for almost the same degree of protection as the patent law. This Utility Model Law of 1905 was clearly intended for the traditional industries and imported small-scale light industries. Its main objective was to encourage the creation and use of new devices among home workers and to promote export of their products.[50]

People reacted to this law enthusiastically. Between July 1905 and 1908, 30 403 applications were submitted and 11 713 utility models were registered. Comparison of Table 8.4 (A) with (B) suggests a far greater relevance of the utility model system to the indigenous inventors.

After the wars with China (1894–5) and Russia (1904–5), the Japanese industrialisation entered its second stage. The emphasis in development effort shifted from light to heavy industries. The proportion of inventions related to traditional industries declined, and the textile industry was gradually replaced by the machinery and chemical industries as the most important areas of invention.[51] The pace of scientific and technological progress in advanced countries was accelerating. To adapt to the changing economic and technological situation, both the patent law and the utility model law were revised in 1909 and 1921.

Throughout the period, the annual rates of patent applications and grants continued to grow. The decline in the 1940s (Table 8.4) is largely attributable to the decreased applications from abroad. The high rates of applications imply that a great majority of inventions took place in the private sector.

The enthusiastic acceptance of the patent and utility model systems by the Meiji Japanese may be attributable partly to historical factors. First, during Edo Era (1605–1868), despite a decree prohibiting inventions except for 'show business', feudal clans encouraged craftsmen to improve production methods and develop new products to sell at national markets of Osaka and Edo (Tokyo), and something similar to the patent system was used for this purpose. After the Meiji Restoration (1868), the craftsmen lost protection from the clans, but became free to plunge into any business and technological venture. What is perhaps most important is that by the mid-1860s the Japanese leaders had become fully convinced of the seriousness of their technological inferiority *vis-à-vis* the Western Powers in view of China's experience in the Opium War and the Satsuma and Nagato clans' bitter defeats in their battles against the Western Powers. Thus, 'Eastern soul and Western science and technology' was a motto to the Meiji Japanese, together with 'Enrich the nation and strengthen the army'. The premature effort at a patent system in 1871 may be interpreted as a sign of the Meiji Government's preoccupation with technology. Similar concern was shared by social reformers like Yukichi Fukuzawa, who enlightened the masses through their lectures and newspaper articles. The literacy rate of the Japanese males is believed to have reached the level of about 40 per cent by the mid-nineteenth century, thanks to the widespread informal education provided by monks and samurais.[52] The primary school enrolment rate of the boys surpassed 50 per cent by 1875, 90 per cent by 1900, and 99 per cent by 1920. Furthermore, primary and secondary education contained an important element of 'ethic training',[53] which was meant to cultivate a sense of obligation to the nation and commitment to the national mottos. Consequently, the urge for technological effort by the government and social leaders was understood and had considerable influence. In this climate of general enthusiasm for nation building, the government introduced a variety of programmes specifically intended to motivate and assist people's effort for technological advancement.

Probably the most important was industrial fairs and competitive exhibitions (Kyôshin-kai). The Emperor personally attended national industrial fairs, and spoke at the opening or commendation ceremony. Those national fairs took place five times between 1877 and 1903, and contributed a great deal to technological development. For example, the Gara spinning machine, which applied a water wheel to traditional manual spinning equipment, spread very rapidly all over the country after it was awarded a special prize as the most valuable invention exhibited at the First National Fair. Another important example is Minorikawa's silk reeling machine. He displayed a four-reeled machine at the national industrial fair of 1903, and a twenty-reeled machine at the Tokyo Industrial Fair of 1907. It provided a technical base for mechanisation and automation of the silk industry.

Modelled after a French system for agricultural development, the competitive exhibitions were intended to encourage export industries (e.g., silk and tea) and promote local industries which were facing competition with imports (e.g., cotton and sugar). Those exhibitions not only encouraged generation of new local technologies and devices, but also provided orientation for future inventions. The products and equipment displayed at these fairs and competitive exhibitions were usually commented upon by referees, who often compared local exhibits with their counterparts in advanced countries. For example, a referee at the competitive exhibition of 1885 argued that it was impossible to beat European yarn while using the local raw cotton of short fibre. Thus, the Osaka Spinning Co., Japan's first successful large-scale modern spinning factory, developed technology to use mixed raw cottons.[54] The practice of mixing the most economic raw cottons imported from different parts of the world is believed to have been one of the secrets of Japan's success in beating Lancashire.

In 1918, the government started an official programme of invention exhibitions, while the Imperial Invention Association and prefectural governments often established an invention museum and awarded prizes to distinguished inventors. In the previous year, the government had started subsidising inventive activity, but the fund for this use remained modest. In 1932 a special committee was established to advise the Minister of Commerce and Industry on the encouragement of invention. The Patent Bureau selected and publicised those inventions which were considered to contain excellent technical ideas, or expected to contribute greatly to the improvement of conditions of life, rationalisation of production, promotion of exports, etc. The Bureau selected 272 such 'remarkable inventions and ideas' between 1940 and early 1945 and publicised them through mass media. The Patent Bureau's extensive information service may be illustrated with reference to the number of copies of official publications on patents and utility models: about 47 000 in 1905, 170 000 in 1910 and over 400 000 by 1927.

A medalling system for inventors was introduced as early as 1881. The Emperor offered bounty for inventors and contributed research funds on many occasions. In 1930 and 1939, he invited ten important inventors to a royal meal.

Non-governmental programmes for encouragement of inventive activity also started early, often gaining the government's and the Emperor's support. With a view to enhancing the popular interest in such activity and enlightening the public regarding the patent system, for example, an Association for the Protection of Industrial Property Right (Kôgyô Shoyûken Hogo Kyôkai) was set up in Tokyo in 1904. It organised conferences and exhibitions of invented items, and offered prizes at invention contests. In 1910, it was renamed the Imperial Invention Association (Teikoku Hatsumei Kyôkai). After 1922, its affiliate associations were set up in different parts of the country to take up similar activities.[55]

The government's programmes aimed at stimulating people's enthusiasm would have failed if people had not had an adequate capacity to react to it. Partly for the purpose of expanding such capacity, the government made tremendous efforts in the domain of education and vocational training.

Pre-war Japan's education programme is characterised by (a) its strong bias towards practical science and (b) the modest scale of higher education and great emphasis on primary and secondary education (Table 8.5).

In 1873, the Ministry of Engineering established an Imperial College of Engineering (Kôbu Daigaku), 'with a view to the education of engineers for service in the Department of Public Works'.[56] Before 1886, 211 students graduated from the College: 48 in mining, 45 in civil engineering, 39 in mechanical engineering, 25 in applied chemistry, 20 in architecture, 8 in shipbuilding and 5 in metallurgy.[57] The subject distribution of the graduates roughly corresponded to the structure of the Ministry's expenditure during its entire period of existence (1870–85). With the abolition of the Ministry, the College lost its *raison d'être* and was absorbed by the Imperial University of Tokyo (former Tokyo University) in 1886.

At Tokyo University, set up in 1877, 90 per cent of the students studied natural sciences including medical science in 1880, and in 1886 the percentage was still over 80. Its faculty of science included

TABLE 8.5 *Universities and Business (agriculture, commerce and industry) Schools in Japan (1875–1945)*

Year	University				Business College				Central govt
	Central govt	Local govt	Private	Total	Central govt	Local govt	Private	Total	
1875									—
1880	1	—	—	1					—
1885	2	—	—	2	2	—	—	2	—
1890	1	—	—	1	3	—	—	2	1
1895	1	—	—	1	3	—	—	3	—
1900	2	—	—	2	4	—	—	4	—
1905	2	—	—	2	10	1	2	13	—
1910	3	—	—	3	13	2	2	17	—
1915	4	—	—	4	17	2	3	22	—
1920	6	2	8	16	20	2	5	27	1
1925	11	4	19	34	44	2	4	50	1
1930	17	5	24	46	42	2	7	51	—
1935	18	2	25	45	44	2	14	60	—
1940	19	2	26	47	51	3	18	72	7
1945	19	2	27	48	61	26	50	137	—

NOTE The length of period of schooling changed from time to time. According to the legislation around 1919,

University: 3–4 years after 13 years' schooling
Business college: 3–4 years after 11 years' schooling
Business school: (A): 3 years after 8 years' schooling

mechanical engineering, civil engineering, mining, metallurgy and applied chemistry. The students were engaged in practical training during the summer vacations. In 1884 shipbuilding was added at the request of the navy.

The objective of the Imperial Universities was 'to teach and study sciences and practical arts in response to the national needs', according to Article 1 of the Imperial University Ordinance of 1886. After Tokyo, imperial universities were established in Kyoto (1887), Kyûshû (1907), and Tôhoku (1911). They started invariably with faculties of natural sciences (science, technology, medical science, and/or agriculture), and faculties of social sciences and humanities (e.g., law) were added later. Here again, the government's preoccupation with technological development is clear.

Around the turn of the century, the Japanese began to realise that business (agriculture, commerce and industry) high schools or colleges were more appropriate means of meeting the nation's manpower requirements than universities. An ordinance of 1899 provided that the authorities of prefectures, cities, towns or villages as well as individual people could establish a business school. A

Business School (A)			Business School (B)				Business Supplementary School			
Local govt	Private	Total	Central govt	Local govt	Private	Total	Central govt	Local govt	Private	Total
—	1	1	—	—	—	—				
7	8	15	—	—	—	—				
19	7	26	—	—	—	—				
19	3	23	—	—	—	—				
37	8	45	1	5	4	10	—	55	—	55
104	12	116	1	19	3	23	1	142	8	151
145	13	158	1	105	8	114	1	2 636	109	2 746
183	21	204	1	266	10	277	1	5 847	263	6 111
183	23	206	1	328	16	345	4	8 578	326	8 908
247	31	279	1	376	20	397	4	14 060	168	14 232
431	96	528	1	247	21	269	4	15 258	54	15 316
577	209	786	1	158	31	190	3	15 193	52	15 248
693	268	961	4	246	42	292	—	16 378	327	16 705
881	319	1 207	4	231	37	272	—	18 280	2 188	20 492
1 245	498	1 743	—	—	—	—	7	12 631	2 415	15 144

Business school: (B): 3 years after 6 years' schooling
Business supplementary school: 3 years after 6 years' schooling

SOURCE Japanese Government, Ministry of Education: *Gakusei 90-nen-shi* (90 years of Japan's official education system) (Tokyo: 1954).

revision of the ordinance in 1920 allowed for different types of industrial education through adaptation in the curriculum and in the length of their courses (between three to five years). Not only chambers of commerce but also agricultural and other trade associations were authorised to set up a business school. The number of business schools grew rapidly (Table 8.5). The content of training at these schools was adapted to the needs of the locality, as their establishment was a result of eager petition from local people.[58] In the meantime, the primary school enrolment rate of boys reached 99 per cent by 1920, as we noted earlier.

The official R&D programme started with the Tokyo Industrial Laboratory which was set up in 1900 as the country's 'central laboratory'.[59] Choosing technologies of nationwide relevance, it undertook research for local production of imported materials (e.g., chemicals), and scientifically analysed traditional local materials and processes (e.g., dyes and ceramics) to find means of their improvement. A similar national laboratory of a general nature was subsequently opened in Osaka, while other national laboratories carried out research in specialised areas such as electrical engineering and railways.

During the boom brought about by the First World War, the official R&D system expanded considerably. A survey in 1931 recorded fifty-six national (as distinguished from prefectural and other local) institutes, excluding seventeen belonging to the army and the navy. Some of the national laboratories obtained considerable numbers of patents and utility models for their inventions. In the case of the two 'central laboratories' in Tokyo and Osaka, however, the major area of contribution is believed to have been related not so much to inventions as to the supporting service provided for individual industrialists' effort for technological progress.[60] For example, they opened their experimenting facilities for public use in 1927 and 1928, respectively. This was intended to help those researchers and industrialists who needed experiments and yet had no access to such facilities elsewhere. The Tokyo Industrial Laboratory had a capacity to accommodate about seventy people.

After 1901 'industrial experiment stations' were set up by local governments to assist industrialists in their technological problem-solving and inventive effort. Located at the level of the prefecture, county or city, they organised lectures, distributed sample products, analysed and tested raw materials, products and machines and tools, provided consultancy service for local producers, and disseminated information concerning results of tests and experiments they had conducted. The number of such stations rose from 14 in 1912 to 60 in 1924, 83 in 1931 and 120 by 1945. A great majority of them were related to textiles including dyeing, followed by brewery, metal, wooden and bamboo handicrafts, lacquer, ceramics, and paper-making.[61]

What emerges from this review of Japan's experience is an image of concerted effort for the stimulation of people's inventive activity by the Emperor, the central and local governments, industrialists and academicians, all of whom enhanced the social status of inventors, kindled people's interest in such activity, reduced constraints on their efforts by providing consultancy, testing and training facilities, and finally helped diffusion of information concerning distinguished and useful inventions. In this respect, the Japanese Government adopted, for the purpose of encouraging indigenous technology development, basically the same tactics as in accelerating the economic development process itself: namely, (a) stimulation of people's initiative; (b) elimination of obstacles to their spontaneous development efforts; and (c) provision of supporting services. Except where risk was too

large for the private sector (introduction of large-scale heavy industries, acquisition and development of totally new large-scale technologies, etc.), the government tried to create an appropriate climate for
people's initiative rather than do it itself. All this appears to contrast
sharply with the situation in most developing countries, where government intervention tends to be direct but limited to one or two
offices' activity without support from others, and where co-operation
between the government on the one hand and industries, academic
circles, etc. on the other is rarely impressive.

SUMMARY AND CONCLUSIONS

No economic development process can spread and sustain itself
without spontaneous participation of the general public. The ever-
growing poverty and unemployment in many parts of the Third
World suggests that, compared with the vastness of the problem,
what a government can directly do is only marginal, especially in a
market economy or mixed economy. Regarding the development of
indigenous technology capacity, this is not simply a matter of establishing and expanding a formal R&D system and programme. It
requires participation of the general public. History indicates that the
patent (including the utility model) system can work as an effective
tool to stimulate people's interest in inventive activity. Third World
nations' frustration about the system appears to accrue partly from
their patent office's failure to provide adequate information services
and the absence of complementary policy instruments, notably anti-
monopoly regulation.

The earlier the stage of industrialisation of a nation, the greater the
proportion of traditional and imported light consumer industries
normally in its industrial sector. Technologies used in these industries
are relatively simple, and inventions associated with such industries
may be of lesser importance compared with those in modern industries and those made by foreign firms, in terms of the value of
industrial output or sales. Economic development of a nation, however, is the result of cumulative effects of innumerable small inventions and improvements. Where people are still unaccustomed to
associating inventive efforts with economic returns, even such 'unimportant' inventions merit encouragement. For this purpose, the utility model system may be more appropriate than the ordinary patent
system.

In modern high technology industries, capacity for patentable invention may be considered to be something which emerges after the development of capacity for repair and maintenance of imported equipment and consumer durables, for their redesigning, and for more substantial adaptation of foreign technology. In this process of sequential evolution of indigenous technological capacity, too, the utility model system can serve as an important stimulus.

To encourage inventive activity, it is necessary to boost people's 'wants to invent' and 'capacity to invent'. The former depends on the return expected from the intended invention and on the awareness of the need for specific types of innovation. Findings in our cross-country analysis in the second section and studies in industrialised countries suggest that expansion of industrial production provides an important stimulus in this respect, although the relationship is, of course, not unidirectional. Industrial competition with other producers at home and/or abroad will oblige industrialists to engage themselves in inventive activity as a means of survival. The experience of Japan, as well as China and the Republic of Korea, suggests that 'wants to invent' can also be boosted through moral suasion which urges individual citizens to contribute to the nation's development. In this connection, it is interesting to note the following remark of Gerschenkron (in a context of French and English industrialisation!): 'in an advanced country rational arguments in favour of industrialisation policies need not be supplemented by a quasi-religious fervour . . . In a backward country the great and sudden industrialisation effort calls for a New Deal in emotions'.[62]

The capacity to invent seems to be fostered, basically, through education and industrial training. Our findings in the second section suggest that primary and secondary education is more relevant to the development of inventive capacity of a nation than higher education. The latter is obviously essential for formal R&D related to modern high technologies. However, a great majority of inventions that are useful for industrial development are problem-solving or improvement inventions, which do not require a formal R&D programme. What is needed is thorough knowledge of products and production processes, capacity to absorb new ideas from reading and observations, and capacity to think. This seems particularly the case in light consumer goods industries which dominate the industrial sector of Third World economies. To encourage this kind of invention, a chain of local industrial experiment stations seems to be more effective than modern R&D institutes copied from industrialised countries.

The R&D and education programmes in many developing countries may need re-examination from this viewpoint.

As regards modern industries, importation of foreign technologies can provide technological opportunities for adaptive inventions. Dependence upon imported technologies need not contradict the long-term goal of technological self-reliance: 'To win time and speed, we must necessarily import advanced technology and equipment from foreign countries. To import them is for the purpose of learning from them and promoting our own creations instead of using them to replace our own'.[63] However, the mode of their importation and application influences the extent of such opportunities significantly, as the NCAER study cited earlier clearly indicates. An important question here relates to attitudes regarding foreign technologies. Without hard commitment to self-reliance, easy access to foreign technologies will kill the nation's enthusiasm for development of indigenous technologies. The current campaign for low-cost transfer of technologies from industrialised countries may or may not benefit developing nations in the long run. On the other hand, one can equally argue that without commitment to self-reliance, the high cost of technology transfer will simply discourage their development effort. Clearly, then, commitment to self-reliance is the key to the whole issue.

APPENDIX: TABLE A.1 *Basic data*

GNPCAPIT	= Gross national product per capita in 1979 in US dollars.
GDMPCAP	= Gross domestic manufacturing production per capita in 1979 in US dollars.
SECON	= Number of people enrolled in secondary school as percentage of age group in 1979.
HIGHER	= Number of people enrolled in higher education as percentage of population aged 20–24 in 1978.
LITRAT	= Adult literacy rate in around 1977, in percentage.
POPAPPL	= Number of inhabitants per patent application in 1978–80.
LGNPCAPI	= Natural logarithm of GNPCAPIT.
LGDMPCAP	= Natural logarithm of GDMPCAP.
LPOPAPPL	= Natural logarithm of POPAPPL.

(-1 and -1.000 indicate non-availability of data).

Country*	Code	(1) GNPCAPIT	(2) GDMPCAP	(3) SECON	(4) HIGHER	(5) LITRAT	(6) POPAPPL	(7) LGNPCAPI	(8) LGDMPCAP	(9) LPOPAPPL
Bangladesh	BA	90	6	25	2	26	2828000	4.500	1.792	14.855
Nepal (1)	NP	130	5	19	3	19	13963000	4.868	1.609	16.452
India (2)	IN	190	30	27	8	36	583707	5.247	3.401	13.277
Malawi	MA	200	25	4	0	25	2908500	5.298	3.219	14.883
Sri Lanka (2)	SL	230	45	53	1	85	454437	5.438	3.807	13.027
Haiti (2)	HT	260	-1	15	1	23	820166	5.561	-1.000	13.617
Pakistan	PK	260	36	16	2	24	2490781	5.561	3.584	14.728
Zaire	ZA	260	8	19	1	58	4584833	5.561	2.079	15.338
Indonesia (2)	ID	370	30	22	3	62	17858736	5.914	3.401	16.698
Zimbabwe (2)	ZW	470	127	15	-1	74	152042	6.153	4.844	11.932
Egypt	EG	480	122	48	15	44	547436	6.174	4.804	13.213
Zambia	ZM	500	92	17	2	44	5580000	6.215	4.522	15.535
Honduras	HN	530	90	21	8	60	712600	6.273	4.500	13.477
Bolivia	BO	550	118	35	13	63	319294	6.310	4.771	12.674
Thailand	TH	590	115	29	7	84	3789583	6.380	4.745	15.148
Philippines	PH	600	150	63	27	75	359600	6.397	5.011	12.793
Nigeria	NG	670	45	13	1	-1	13767166	6.507	3.807	16.438
El Salvador	ES	670	119	26	8	62	275625	6.507	4.779	12.527
Peru (2)	PE	730	223	50	16	80	176793	6.593	5.407	12.083
Morocco	MO	740	130	22	4	28	723629	6.607	4.868	13.492
Colombia	CO	1010	202	46	10	-1	555787	6.918	5.308	13.228
Syrian Arab Rep (2)	SY	1030	221	47	18	58	1439833	6.937	5.398	14.180
Ecuador	EC	1050	223	49	35	81	269366	6.957	5.407	12.504
Tunisia	TU	1120	117	25	5	62	326000	7.021	4.762	12.695
Jamaica	JM	1260	166	58	-1	90	270250	7.139	5.112	12.507
Turkey	TR	1330	268	34	8	60	451397	7.193	5.591	13.020
Panama (2)	PA	1400	-1	66	20	-1	149500	7.244	-1.000	11.915
Korea (Republic of)	KO	1480	433	76	12	93	34755	7.300	6.071	10.456
Mexico (2)	ME	1640	537	45	12	81	96620	7.402	6.286	11.479

Country	Code									
Chile (2)	CL	1690	459	55	12	-1	80272	7.432	6.129	11.293
South Africa	SA	1720	408	-1	-1	-1	9940	7.450	6.011	9.204
Brazil (2)	BE	1780	491	32	11	76	56737	7.484	6.196	10.946
Costa Rica	CR	1820	350	48	24	90	74551	7.507	5.858	11.219
Uruguay	UR	2100	646	59	24	94	55846	7.650	6.471	10.930
Iran (Islamic Rep of)	IR	-1	-1	44	18	50	303040	-1.000	-1.000	12.622
Portugal	PT	2180	702	55	5	70	95794	7.687	6.554	11.470
Argentina	AR	2230	1288	56	11	93	19849	7.710	7.161	9.896
Iraq	IQ	2410	145	56	22	-1	451107	7.787	4.977	13.019
Yugoslavia	YU	2430	861	82	9	85	17768	7.796	6.758	9.785
Venezuela	VE	3120	542	40	23	82	69821	8.046	6.295	11.154
Hong Kong	HK	3760	665	63	21	90	248250	8.232	6.500	12.422
Greece	GR	3960	683	81	11	-1	6515	8.284	6.526	8.782
Israel	IS	4150	-1	68	18	-1	5985	8.331	6.877	8.697
Ireland	IL	4210	970	92	26	98	9114	8.345	-1.000	9.118
Spain	SP	4380	-1	78	19	-1	19942	8.385	-1.000	9.901
New Zealand	NZ	5930	1302	81	24	99	2908	8.688	7.172	7.975
United Kingdom	UK	6320	1794	83	29	99	2870	8.751	7.492	7.962
Finland	FI	8160	2226	90	20	100	3565	9.007	7.708	8.179
Austria	AT	8630	2647	72	21	99	3176	9.063	7.881	8.063
Japan	JP	8810	2525	90	22	99	758	9.084	7.834	6.631
Australia	AU	9120	-1	86	29	100	2705	9.118	-1.000	7.903
Canada	CA	9640	1820	89	26	99	13513	9.174	7.507	9.511
France	FR	9950	2675	84	37	99	4745	9.205	7.892	8.465
Netherlands	NL	10230	3086	93	24	99	7139	9.233	8.035	8.873
United States	US	10630	2522	97	28	99	3644	9.271	7.833	8.201
Norway	NO	10700	2389	94	56	99	5342	9.278	7.779	8.583
Belgium	BE	10920	2934	86	25	99	10848	9.298	7.984	9.292
Germany (Fed. Rep.)	GE	11730	4746	94	26	99	2041	9.370	8.465	7.621
Denmark	DK	11900	-1	83	26	99	5484	9.384	-1.000	8.610
Sweden	SE	11930	2824	86	29	99	1940	9.387	7.946	7.570
Switzerland	SW	13920	-1	55	37	99	1485	9.541	-1.000	7.303

NOTES   *The figures in the parentheses in this column indicate that the number of inhabitants per patent application for the country is not an average of three years, but one of two years or relates to a single year. In the case of Peru, no application was recorded in 1979 either for residents or for non-residents. In the light of her records before and after this year, one cannot but suspect that there must have been some unusual explanation. We therefore took the average of 1978 and 1980, instead of counting 'zero' for 1979.

SOURCES   Column (1): The World Bank: *World Development Report 1981*.

Column (2): Calculated using data on the GDP in *World Development Report 1981* and population statistics in *1981 World Bank Atlas*. In the case of Nepal and Syria, the percentage share of manufacturing sector for 1980, the percentage share of the manufacturing sector for 1980 published in the *World Development Report 1982* was used because such figure for 1979 is not available.

Columns (3), (4) and (5): *World Development Report 1982*.

Column (6): Annual average of patent applications for 1978–1980 in WIPO: *Industrial Property Statistics* divided by 1979 population figures in *1981 World Bank Atlas*.

NOTES

1. In finishing this study, I have benefited a great deal from comments from Hamid Tabatabai, Ajit S. Bhalla, Shigeru Ishikawa, Jeffrey James and Larry Westphal. My thanks are also due to Tetsuo Tomita, Hideo Nakamura and Nobuo Suzuki of the Japanese Patent Office, and to Tsutomu Hosaka of the Japan Institute of Invention and Innovation, who kindly took the trouble of reading an early draft, pointed out a number of factual mistakes, and offered useful advice. The computer work related to the second section of this chapter was done by Josiane Capt. I am, however, entirely responsible for whatever faults may remain.

2. Usually, a 'patent' describes an invention and creates, for a limited time, a legal situation in which the patented invention can normally be exploited only with the authorisation of the owner, while the exploitation of the invention requires the authorisation of the state where an inventors' certificate is involved. A 'utility model' ('certificat d' utilité' in France) usually differs from a 'patent' in the following respects: the former is not subject to examination as to substance and does not require non-obviousness; the duration of the protection is shorter and sometimes it is available only for mechanical inventions (WIPO, 1982). In reality, however, the usage of these terms varies considerably from one country to another. For example, a utility model in Japan is issued after examination, while a patent in France is issued without examination. We use the concept of 'utility model' in the sense it is applied in Japan, i.e., a patent for an invention which is simpler than one for which an ordinary patent is issued (cf. fifth section below).

3. Cunningham (1915–21), vol. II, p. 75. Foreign patentees were to instruct English apprentices in the practice of new arts so that at the end of the term of their patents the arts could be substantially used by Englishmen (ibid., pp. 53–84). Cecil's success with the patent system seems to have been due partly to the fact that the introduction of this system coincided with the inflow of Protestant refugees from the Continent. Many of the beneficiaries, however, were Englishmen: only 21 out of the total 55 patents granted between 1561 and 1603 were for foreigners (Hulme, 1900, p. 52).

4. This survey conducted in 1979 and 1980 involved 2390 firms. Out of the total, 711 (29.7 per cent) mentioned the patent system as the most important incentive; 322 (13.5 per cent) tax incentive; 302 (12.6 per cent) other financial incentives, and 207 (8.7 per cent) a variety of commendation systems. Regarding the motivation of researchers in those firms, the most important was competition with other firms (706 out of 3052 respondents, or 22.8 per cent), followed by academic or technical interest (522 or 16.8 per cent), patents (539 or 11.6 per cent), and targets set by the firm (343, or 11.1 per cent) (Hatsumei Kyôkai, 1981, pp. 127 and 147).

5. See, for example, Vaitsos (1972), Patel (1974), Roffe (1974), and UNCTAD (1974 and 1981). Penrose (1973) takes a more moderate view.

6. Greer, (1973), for example.
7. Cf. Watanabe (1983), chs. II and IX, and Rosenberg (1963). Today's microelectronics-based technological revolution has originated from defence industries. For potential civil application of technologies developed under the NASA programme of the United States, see UNIDO (1982). A great variety of 'new materials' useful in civil industries have also been developed in connection with this programme (*Nihon Keizai Shimbun* (Tokyo), 1 May 1983, morning).
8. Cf. Lall (1982).
9. Indian Government, Planning Commission: *Sixth Five Year Plan 1980–1985*, p. 319.
10. An eminent exception is Herrera (1972).
11. Strassmann (1968), p. 50.
12. For example, many patents are taken without any productive use in view. A single patent or even a cluster of patents may be useless to the patentee if some complementary patents and know-how are unaccessible. Patents may be taken simply to control the market for the products and not to use them. Also a considerable amount of developmental work may be needed before the commercial application of patented inventions can take place.
13. The propensity to patent an invention of given quality varies from firm to firm and from industry to industry, and the quality of underlying invention varies widely from patent to patent (cf. Scherer, 1965, p. 1098; and Schmookler 1966, ch. II). See also Kamien and Schwartz (1975), pp. 4–5.
14. Differences among the industrialised countries are probably not very significant, considering the limited comparability of data. Patent statistics in some countries include renewals of old patents, 'patents in addition', etc. which may not be included in other countries. Moreover, firms which make inventions under government contract (e.g., aircraft and other armament manufacturers) seldom seek patent protection, since they must give the government either exclusive rights or at least royalty-free license in any event (Scherer, 1965, p. 1101). In 1977 the government's share in the national R&D expenditure excluding R&D related to defence was 27 per cent in Japan, 41 per cent in the Federal Republic of Germany, about 51 per cent in the United States, and 53 per cent in France (Japanese Government, Science and Technology Agency,1981, pp. 136–7).
15. Kennedy and Thirlwall (1972), p. 52.
16. Sangyô Kenkyû-jo (1981), p. 283.
17. Schmookler (1966).
18. Japanese Government, Patent Office (1955), pp. 140–1.
19. Scherer (1965 and 1982).
20. Rosenberg (1974).
21. Schmookler and Brownlee (1962), pp. 166–7.
22. Japanese Government, Science and Technology Agency (1981), p. 137.
23. Watanabe (1981).
24. Indian Government; *Research and Development Statistics 1976–77*, p. 69.

25. Soundararajan (1977), pp. 70–1.
26. Data kindly supplied by NRDC during the author's visit to India in November–December 1980.
27. Annual Report of the NRDC quoted in Desai (1980), pp. 90–1.
28. Indian Government, CSIR (1979), p. 2.
29. Indian Government, NCAER (1971), pp. 49–51.
30. More recently, a senior chemical engineer of the NRDC elaborated somewhat on the causes of the problem (Soundararajan (1977)).
31. Bagchi (1980), p. 307.
32. Ahmad (1978), p. 2085.
33. Desai (1980), pp. 90–2.
34. Thomas and Wionczek (1979), p. 231.
35. Cf. Chudnovsky (1983), p. 189. Patent protection tends to increase the cost of imitation in the pharmaceutical industry much more than in other industries (see Mansfield, Schwartz and Wagner (1981) for some evidence).
36. Morehouse, Gupta and Deolalikar (1980).
37. UNCTAD (1981), p. 21.
38. Pai (1956).
39. Aboagye (1982).
40. Little, Scitovsky and Scott (1970), Bhagwati and Desai (1970) and Das (1972).
41. Morehouse, Gupta and Deolalikar (1980).
42. Vaitsos (1972), p. 79.
43. Japanese Government, Science and Technology Agency (1981).
44. Emi (1963), Umetani (1965) and Miyoshi (1979).
45. Cf. Myrdal (1968), pp. 1647–8 for Asia, and Thomas and Wionczek (1979) and Street and James (eds) (1979) for the Caribbean and Latin America.
46. Indian Government, Planning Commission: *Sixth Five Year Plan 1980–85*, p. 318.
47. Takahashi (1976), vol. I, p. 190. However, examples of important inventions are not missing in those years. Jinrikisha (rickshaw) was originally invented around 1869, and by the end of 1871 24 500 units were in operation in Tokyo and its periphery. Its export to China, France and other countries started after 1875. In 1871 a kind of cotton flannelette was invented by combining the elements of local crepe with those of German flannel. This was first adopted as material of army uniforms and later exported in large quantities. The Gara spinning machine, which applied a water wheel to the traditional local spinning equipment for semi-mechanisation purpose, is another notable example. In the wake of the practice during the feudal period, the government was kept informed of inventions. Between 1873 and 1884, the Ministry of Engineering (Kôbu-shô) and the Ministry of Agriculture and Commerce registered 80 inventions including 16 agricultural and fishing implements, 15 textile machines and 4 dyeing techniques (Japanese Government, MITI (1964), pp. 88–91). In the process of preparation of a centenary of the Japanese patent system, a group of the Japanese Patent Office staff has been discovering an increasing number of inventions reported to local governments.

48. Ibid., pp. 130–1.
49. Under the threat of Western battleships, and in the light of China's bitter experience of the Opium War (1839–42), Japan accepted treaties which denied her tariff autonomy and provided for extra-territorial rights for Europeans (including Russians) and Americans. Revision of those 'unequal treaties' was completed only in 1911, when tariff autonomy was recovered.
50. Japanese Government, MITI (1964), pp. 142–6.
51. Japanese Government, Patent Office (1955), pp. 178–9.
52. Dore (1965).
53. Fridell (1970).
54. Japanese Government, MITI (1979), pp. 30–9 and 75–6.
55. Japanese Government, Patent Office (1955), pp. 434–43 and MITI (1964).
56. This is a direct citation from the English calendar prepared by Henry Dyer, principal of the College, and printed in 1873, which is quoted in Miyoshi (1979), p. 274. Dyer called 'Kôbu-shô' 'the Department of Public Works', and 'Kôbu Daigaku' 'Imperial College of Engineering'. There is no official English name of 'Kôbu-shô'. Professor Emi, for example, calls it 'Ministry of Industry' (Emi (1963)). Considering the composition of work done by the Ministry during its entire period of existence, I believe that 'Ministry of Engineering' is more appropriate.
57. Miyoshi (1979), pp. 298–9. For the development of the secondary level industrial training schools in pre-war Japan, see also Tokyo Kôgyô Daigaku (1940).
58. Japanese Government, MITI (1979), p. 164.
59. On the Tokyo Industrial Laboratory, see Tokyo Kôgyô Shikenjo (1951).

In the private sector, the leader was the mining industry, where Sumitomo with Besshi Mine established a metallurgical research institute in 1880, followed by Mitsui (1893) and Mitsubishi (1907). The first laboratory in the electrical machinery industry was opened by Shibaura in 1899. In metal engineering, Mitsubishi Shipbuilding started an analysis office in 1904 (cf. Kamatani, 1965, pp. 401–2). The number of research institutes in the private sector increased from 6 in 1902 to 34 in 1912, 162 in 1923, and 193 in 1931.

The history of Rikagaku Kenkyûjo (RIKEN) illustrates the enormous enthusiasm which Japanese industrialists and scholars showed for inventive activity. Established jointly by the government and the private sector in 1917, it was meant to consolidate the base for 'enriching the nation and strengthening the army' through the attainment of technological self-reliance and through the development of a strong national supply base with regard to advanced technologies. Its staff was divided into more than two dozen groups, which were headed by celebrated professors from different universities. They competed with each other in terms of the number of prizes and patents and the amount of profits resulting from the commercial application of their research results. RIKEN launched companies to put their newly developed technologies to commercial use. Those companies formed the so-called 'RIKEN Konzern'. By 1937, over 60 per cent of the revenue of the institute came from its member companies, and RIKEN declined the government's

254   *The Patent System and Technology in the Third World*

financial contribution in the same year (Japanese Government, MITI, 1979, pp. 512–13 and 518).
60. Ibid., p. 518.
61. Japanese Government, MITI (1979), p. 242. Many localities in Japan are still known for particular types of such traditional industries which have centuries' history (cf. Yamazaki (1980)).
62. Gerschenkron (1962), pp. 24–5.
63. Deng Xiaopin, quoted in *The China Business Review* (Washington, DC), Sept–Oct, 1977, p. 11.

REFERENCES

Aboagye, A.A. (1982) 'Technology and Employment in the Capital Goods Industry in Ghana', mimeographed World Employment Programme research working paper, restricted (Geneva: ILO), February.
Ahmad, Aqueil (1978) 'Science and Technology in Development: Policy Options for India and China', *Economic and Political Weekly* (Bombay), 23–30 December.
Bagchi, Amiya Kumar (1980) 'Formulating a Science and Technology Policy; What Do We Know About Third World Countries', *Economic and Political Weekly*, Annual Number, February.
Bhagwati, Jagdish N. and Desai, Padma (1970) *India, Planning for Industrialisation – Industrialisation and Trade Policies Since 1951* (London: Oxford University Press).
Chudnovsky, Daniel (1983) 'Patents and Trademarks in Pharmaceuticals', *World Development*, March.
Cunningham, W. (1915–21) *The Growth of English Industry and Commerce in Modern Times*, 3 vols. (Cambridge University Press).
Das, Nabagopa (1972) *The Indian Economy Under Planning* (Calcutta: World Press Private Limited).
Dore, R.P. (1965) *Education in Tokugawa Japan* (London: Routledge and Kegan Paul).
Desai, Ashok V. (1980) 'The Origin and Direction of Industrial R&D in India', *Research Policy*, No. 9.
Emi, Koichi (1963) *Government Fiscal Activity and Economic Growth in Japan 1868–1960* (Tokyo: Kinokuniya Bookstore).
Fridell, Wilvur M. (1970) 'Government Ethics Text-Books in Late Meiji Japan', *Journal of Asian Studies*, August.
Gerschenkron, Alexander (1962) *Economic Backwardness in Historical Perspective: A Book of Essays* (Cambridge, Mass: Harvard University Press).
Greer, Douglas (1973) 'The Case Against Patent Systems in Less-Developed Countries', *Journal of International Law and Economics*, December.
Hatsumei Kyôkai (Japan Institute of Invention and Innovation) (1981) *Gijutsu Kakushin to Tokkyo Seido* (Technological Innovations and the Patent System), Tokyo, March.
Herrera, Amilcar (1972) 'Social Determinants of Science Policy in Latin America: Explicit Science Policy and Implicit Science Policy', *Journal of Development Studies*, October.

Hulme, E. Wyndham (1896) 'The History of the Patent System Under the Prerogative and at Common Law', *Law Quarterly Review*, April.

Hulme, E. Wyndham (1900) 'The History of the Patent System Under the Prerogative and at Common Law, A Sequel', *Law Quarterly Review*, January.

Indian Government, Patent Office: *Annual Report* for various years.

Indian Government (1956) *Indian Patents Centenary (1856–1956), Souvenir* (Calcutta).

Indian Government, National Council of Applied Economic Research (NCAER) (1971) *Foreign Technology and Investment: A Study of Their Role in India's Industrialisation* (New Delhi).

Indian Government, Council of Scientific and Industrial Research (CSIR) (1979) *Proceedings of the 79th Meeting of the Governing Body of the Council of Scientific and Industrial Research*, dated 20 December.

Indian Government, Department of Science and Technology: *Research and Development Statistics 1976–77* (New Delhi).

Indian Government, Planning Commission: *Sixth Five Year Plan 1980–85* (New Delhi).

Japanese Government, Tokkyo-chô (Patent Office): *Tokkyo-chô Nempô* (Annual Report) (Tokyo), various years.

Japanese Government, (1955) *Tokkyo Seido 70-nen-shi* (A 70-Year History of Japan's Patent System) (Tokyo: Hatsumei Kyôkai).

Japanese Government, Mombu-shô (Ministry of Education) (1964) *Gakusei 90-nen-shi* (A 90-Year History of Japan's Official Education System) (Tokyo).

Japanese Government, Tsûshô Sangyo-shô (Ministry of International Trade and Industry (MITI)) (1964) *Shô-kô Seisaku-shi, vol. 14; Tokkyo* (History of Industrial and Trade Policy, vol. 14: Patent) (Tokyo: Shôkô Seisaku-shi Kankô-Kai).

Japanese Government, (1979) *Shô-kô Seisaku-shi, vol. 13: Kôgyô Gijutsu* (History of Industrial and Trade Policy, vol. 13; Industrial Technology) (Tokyo).

Japanese Government, Kagaku Gijutsu-chô (Science and Technology Agency) (1981) *Kagaku Gijutsu Hakusho* (White Paper on Science and Technology) for 1981, (Tokyo).

Kamatani, Chikayoshi (1965) 'The Role Played by the Industrial World in the Progress of Japanese Science and Technology', *Journal of World History*, vol. IX, no. 2.

Kamien, Morton I. and Schwartz, Nancy L., (1975) 'Market Structure and Innovation: A Survey', *Journal of Economic Literature*, March.

Kennedy, Charles and Thirlwall, A.P. (1972) 'Technical Progress, A Survey of Literature', *Economic Journal*, March.

Lall, Sanjaya (1982) *Developing Countries as Exporters of Technology* (London: Macmillan).

Little, I., Scitovsky, T., and Scott, M. (1970) *Industry and Trade in Some Developing Countries, A Comparative Study* (London: Oxford University Press).

Mansfield, E., Schwartz, M., and Wagner, S. (1981) 'Imitation Costs and Patents; An Empirical Study', *The Economic Journal*, December, pp. 907–18.

Miyoshi, Nobuhiro (1979) *Nihon Kôgyô Kyôiku Seiritsu-shi no Kenkyû, Kindai Nihon no Kôgyôka to Kyôiku* (A Study on the History of Industrial Education in Japan; Industrialisation and Education in Modern Japan) (Tokyo: Kazama Shobô).

Morehouse, W., Gupta, B. K. and Deolalikar, A. (1980) *Assessment of US-Indian Science and Technology Relations; An Analytical Study of Past Performance and Future Prospects: The Political Economy of Science and Technology in North–South Relations* (Springfield Virginia: National Technical Information Service), mimeo-typed preliminary book manuscript.

Myrdal, Gunnar (1968) *Asian Drama* (New York: Pantheon), 3 vols.

Pai, K. Rama 'Is There a Future For the Patent System in India?', in Indian Government, Patent Office (1956), above.

Patel, Surendra J. (1974) 'The Patent System and the Third World', *World Development*, September.

Penrose, Edith (1973) 'International Patenting and the Less-Developed Countries', *Economic Journal*, September.

Roffe, Pedro (1974) 'Abuses of Patent Monopoly: A Legal Appraisal', *World Development*, September.

Rosenberg, Nathan (1963) 'Technological Change in the Machine Tool Industry, 1840–1910', *Journal of Economic History*, December.

Rosenberg, Nathan (1974) 'Science, Invention and Economic Growth', *Economic Journal*, March.

Sangyô Kenkyûjo (Institute of Industrial Research) (1981) *Shin Gijutsu Kaihatsu ni okeru Jitsuyô Shin-an Seido no Yakuwari* (The Role of the Utility Model System in New Technology Development) (Tokyo).

Scherer, F.M. (1965) 'Firm Size, Market Structure, Opportunity and the Output of Patented Inventions', *American Economic Review*, September.

Scherer, F.M. (1982) 'Demand-Pull and Technological Invention: Schmookler Revisited', *Journal of Industrial Economics*, March.

Schmookler, Jacob and Brownlee, Oswald (1962) 'Determinants of Inventive Activity,' *American Economic Review, Papers and Proceedings*, May.

Schmookler, Jacob (1966) *Invention and Economic Growth* (Cambridge, Mass: Harvard University Press).

Soundararajan, P. (1977) 'Development and Transfer of Indigenous Technology: Problems and Prospects', *Small Industry Bulletin for Asia and the Pacific* (UN), no. 14.

Strassmann, W. Paul (1968) *Technological Change and Economic Development: The Manufacturing Experience of Mexico and Puerto Rico* (Ithaca, NY: Cornell University Press).

Street, James H. and James, Dilmus D. (eds) (1979) *Technological Progress in Latin America: The Prospects for Overcoming Dependency* (Boulder, Colorado: Westview Press).

Takahashi, Korekiyo (1976) *Takahashi Korekiyo Jiden* (Autobiography), 1936, a modern edition (Tokyo, Chûô Kôron-sha), 2 vols.

Thomas, D. Babatunde and Wionczek, Miguel S. (eds) (1979) *Integration of Science and Technology With Development* (New York: Pergamon Press).

Tokyo Kôgyô Daigaku (Tokyo Institute of Technology) (1940) *Tokyo Kôgyô Daigaku 60-Nen-shi* (A 60-Year' History of Tokyo Institute of Technology) (Tokyo).

Tokyo Kôgyô Shikenjo (Tokyo Industrial Laboratory) (1951) *Tokyo Kôgyô Shikenjo 50-nen-shi* (50 Years History of Tokyo Industrial Laboratory) (Tokyo).

Umetani, Noboru (1965) *Oyatoi Gaikokujin* (Hired Foreigners) (Tokyo: Nihon Keizai Shimbun-sha).

UNCTAD (1974) *The Role of the Patent System in the Transfer of Technology to Developing Countries*, a mimeographed document (TD/B/AC.11/19), April.

UNCTAD (1981) *Review of Recent Trends in Patents in Developing Countries*, mimeographed document (TD/B/C.6/AC.5/3), 24 November.

UNIDO (1982) *Potential Applications of Space-Related Technologies to Developing Countries*, prepared for the Second United Nations Conference on the Exploration and Peaceful Uses of Outer Space, a mimeographed document (A/CONF.101/BP/IGO/13), 26 July.

Vaitsos, C.V. (1972) 'Patents Revisited; Their Function in Developing Countries', *Journal of Development Studies*, October.

Watanabe, Susumu (1981) 'Multinational Enterprises, Employment and Technology Adaptations', *International Labour Review*, November–December.

Watanabe, Susumu (ed.) (1983) *Technology, Marketing and Industrialisation; Linkages Between Large and Small Enterprises* (New Delhi: Macmillan).

WIPO *Industrial Property Statistics, 1979* (Geneva), various years.

WIPO (1982) 'Draft Joint Inventive Activity Guide', prepared for the Committee of Experts on Joint Inventive Activity, Geneva 2–6 May (Doc. No. JIA/II/2).

World Bank *World Development Report* (Washington, DC), various years.

World Bank (1981) *World Bank Atlas*, 1981.

Author Index

n = note; q = quote; r = reference; t = table

258

Subject Index